797,885 Books
are available to read at

Forgotten Books

www.ForgottenBooks.com

Forgotten Books' App
Available for mobile, tablet & eReader

ISBN 978-1-330-50365-2
PIBN 10070726

This book is a reproduction of an important historical work. Forgotten Books uses state-of-the-art technology to digitally reconstruct the work, preserving the original format whilst repairing imperfections present in the aged copy. In rare cases, an imperfection in the original, such as a blemish or missing page, may be replicated in our edition. We do, however, repair the vast majority of imperfections successfully; any imperfections that remain are intentionally left to preserve the state of such historical works.

Forgotten Books is a registered trademark of FB &c Ltd.
Copyright © 2015 FB &c Ltd.
FB &c Ltd, Dalton House, 60 Windsor Avenue, London, SW19 2RR.
Company number 08720141. Registered in England and Wales.

For support please visit www.forgottenbooks.com

1 MONTH OF FREE READING

at

www.ForgottenBooks.com

By purchasing this book you are eligible for one month membership to ForgottenBooks.com, giving you unlimited access to our entire collection of over 700,000 titles via our web site and mobile apps.

To claim your free month visit:

www.forgottenbooks.com/free70726

* Offer is valid for 45 days from date of purchase. Terms and conditions apply.

English
Français
Deutsche
Italiano
Español
Português

www.forgottenbooks.com

Mythology Photography **Fiction**
Fishing Christianity **Art** Cooking
Essays Buddhism Freemasonry
Medicine **Biology** Music **Ancient Egypt** Evolution Carpentry Physics
Dance Geology **Mathematics** Fitness
Shakespeare **Folklore** Yoga Marketing
Confidence Immortality Biographies
Poetry **Psychology** Witchcraft
Electronics Chemistry History **Law**
Accounting **Philosophy** Anthropology
Alchemy Drama Quantum Mechanics
Atheism Sexual Health **Ancient History**
Entrepreneurship Languages Sport
Paleontology Needlework Islam
Metaphysics Investment Archaeology
Parenting Statistics Criminology
Motivational

RUHMKORFF
INDUCTION-COILS,

THEIR

Construction, Operation and Application.

WITH CHAPTERS ON

BATTERIES, TESLA COILS AND ROENTGEN RADIOGRAPHY

BY

H. S. NORRIE.

NEW YORK:

SPON & CHAMBERLAIN, 12 CORTLANDT ST.

LONDON:

E. & F. N. SPON, 125 STRAND.

1896.

Entered, according to Law, with the Librarian of Congress, Washington, D. C.,

By SPON & CHAMBERLAIN, 1896.

Burr Printing House, 18 Jacob Street, New York, U.S.A.

PREFACE.

AT the present time, when so many startling and important phenomena have been produced by high-tension currents from the coils of Ruhmkorff and Tesla, there appears to be an opening for a practical handbook on such.

The intent of the following pages is to give in simple language to the reader such practical information on Ruhmkorff and Tesla coils as will enable him to construct and readily operate them, at the same time avoiding wherever not absolutely imperative any discussion of abstruse electrical theories.

In the chapter on Roentgen photography the writer has referred to the concise and lucid articles on the latter which have iv *Preface.*

appeared from time to time in those papers invaluable to the electrician, the *Electrical World, Electrical Engineer,* and *Western Electrician.*

CONTENTS OF CHAPTERS.

Chapter   I. Coil Construction.
"   II. Contact Breakers.
"   III. Insulations and Cements.
    IV. Condensers.
    V. Experiments.
    VI. Spectrum Analysis.
"   VII. Currents in Vacuo.
"   VIII. Rotating Effects.
    IX. Gas Lighting and Ozone Production.
    X. Primary Batteries.
    XI. Storage Batteries and Electric Light Currents.
"   XII. Tesla and Hertz Effects.
"   XIII. Roentgen Rays and Radiography.

LIST OF ILLUSTRATIONS.

Fig.		Page
1.	Section of Coil	4
2.	Insulating Tube Ends	10
3.	Sectional Winding	11
4.	Section " First Method	12
5.	" " Second Method	13
6.	Proportional Diagram of Coil	15
7.	Section Winder, End View	16
8.	" " Face "	16
9.	Assembly of Coils	18
10.	Polechanging Switch	31
11.	Contact Breaker, Simple	34
12.	" " Imperfect Form	36
13.	" " Superior "	36
14.	Spottiswoode Breaker	38
15.	Polechanging "	43
16.	Leyden Jar	56
17.	Plate Condenser	57
18.	Arrangement of Condenser Plates	59
19.	Condenser Charging, First Method	65
20.	" " Second Method	67
21.	Spark between Balls	75
22.	Short Spark between Balls	75
23.	Sparkling Pane	75
24.	Luminous Design	78
25.	Electric Brush	78

List of Illustrations.

Fig.		Page
26.	Spectrum—Solar	82
27.	Spectroscope and Coil	83
28.	Simple Air Pump	91
29.	Geissler " "	94
30.	Sprengel " "	94
31.	Solution Tube	100
32.	Fluorescent Bulbs	100
33.	Ruby Tube—Crookes	100
34.	Iridio-platinum Tube—Crookes	101
35.	Revolving Wheel	104
36.	Tube Holder	107
37.	Side View of Wheel	107
38.	Geissler Tubes	110
39.	Triangle on Disc	111
40.	Maltese Cross on Disc	111
41.	Gas Lighting Circuit	115
42.	Ozone	118
43.	Grenet Cell	122
44.	Fuller "	126
45.	Gethins Cell	135
46.	Lead Plate	138
47.	Wooden Separator	138
48.	Charging with Rheostat	143
49.	" " Lamps	143
50.	Hydrometer	147
51.	Hertz Resonator	158
52.	Tesla Circuit	160
53.	" Cut Out	162
54.	" " " Top Plan	163
55.	" Coil	164
56.	Crookes Tube	171
57.	Roentgen Circuit	173

CHAPTER I.

COIL CONSTRUCTION.

IN commencing a description of the Ruhmkorff coil and its uses, a brief mention of the fundamental laws of induetion directly bearing on its action will assist in obtaining. an intelligent conception of the proper manner in which it should be constructed and handled.

Any variation or cessation of a current of electricity flowing in one conductor will induce a momentary current in an adjacent conductor; and if the second conductor be an insulated wire coiled around the first conductor, also a coil of insulated wire, the effect is heightened. The intensity of the secondary or induced current increases with the number of turns of its conductor, the abruptness and completeness of the variation of current in **the** first or primary coil, and the proximity of the coils. And the insertion of a mass of soft iron within the primary coil by its consequent magnetization and demagnetization augments still further the inductive effect. There are other contributing causes which cannot be treated of here, but are of not so much importance as the foregoing.

In the Ruhmkorff coil, which is an application of the above laws, the primary coil is of large wire and the secondary coil of extremely fine wire, of a length many thousand times greater than the wire of the primary coil. The current is abruptly broken in the primary circuit by a suitable device—the contact breaker or rheotome. The current induced in the secondary at the make of the circuit is in the opposite direction to that of the primary coil and battery, but the current at the break of the circuit is in the same direction as that of the primary. The effect of the current at the break of the circuit is more powerful than that at the make, which latter is also somewhat neutralized by the opposing battery current. A condenser or Leyden jar is connected across the contact breaker to absorb an *extra current* induced in the primary coil by the break of the circuit, which would tend to prolong the magnetization of the core beyond the desired limit.

The whole apparatus is mounted on a wood base, having the condenser in a false bottom for the sake of compactness.

It is not herein intended to describe all the minor operations in the construction of a Ruhmkorff coil. A sufficient description and review of the main points to be considered, however, will be given to enable a person fairly proficient in the use of simple tools to construct a serviceable instrument.

The parts and their arrangement in relation to one another are shown in Fig. 1, but are not drawn strictly to scale, although very nearly so.

Coil Construction.

C is the core, consisting of a bundle of soft iron wires as fine as can be obtained.

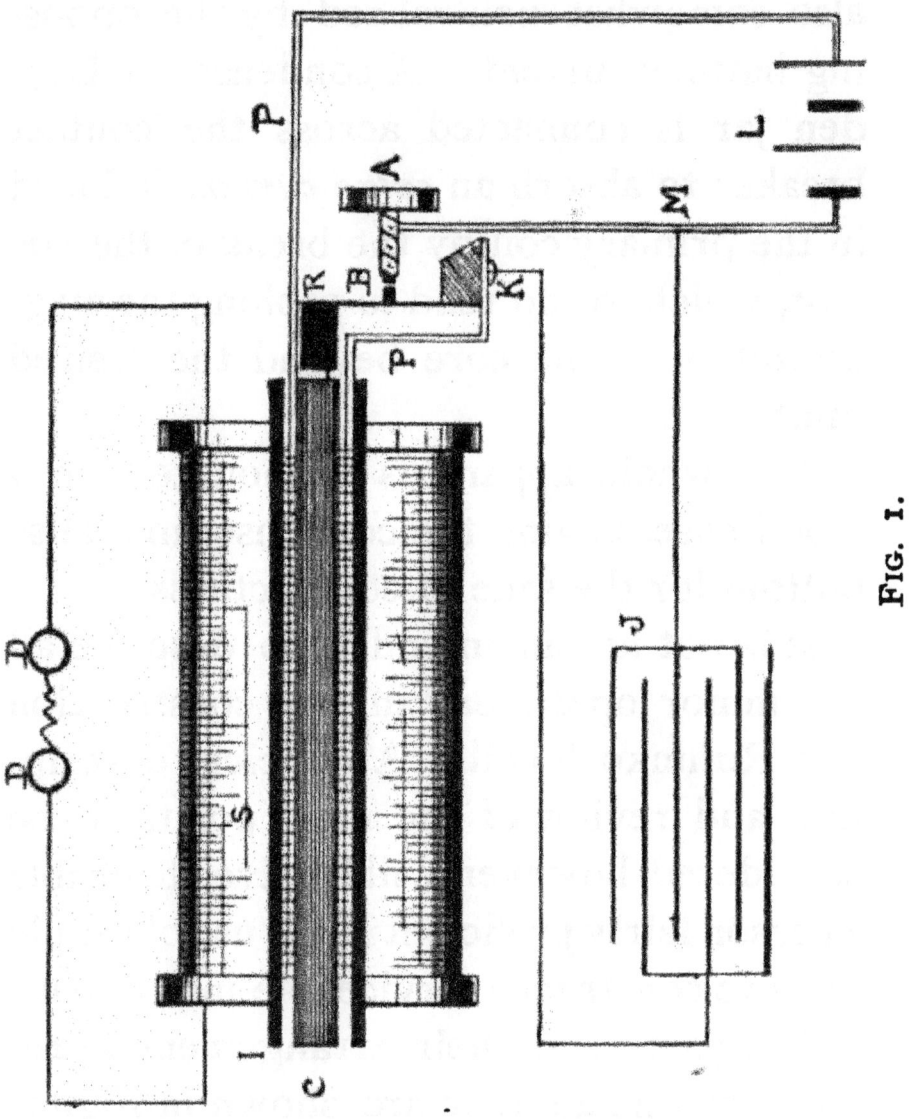

FIG. 1.

The greater the subdivision of the core the quicker will it respond to the magnetizing current in the primary coil, and lose its magnetism when the current ceases. It has another advantage, in that the disadvantageous eddy, or Foucault currents, are lessened, which fact, however, is of not enough importance to need extended consideration.

Many coil-makers saturate the core with paraffin or shellac, which is of slight benefit. This core is wrapped in an insulating layer of paraffined paper or enclosed in a rubber shell, there not being any great necessity to use more than ordinary insulation between the core and the primary coil.

In the majority of induction coils or "transformers" used in the alternating current system of electric lighting, the iron cores form a closed magnetic circuit. A closed magnetic circuit in a Ruhmkorff coil could be obtained by extending the iron core at each end and then bending and securing the ends together, forming, as it were, a ring partly inside and partly outside the coil. But although the inductive effects would be heightened and less battery power required, the slowness of the circuit to demagnetize would alone be detrimental to rapid oscillations of current.

There would also be a loss from a greater hysteresis (energy lost in the magnetization and demagnetization of iron). A core magnetizes quicker than it demagnetizes, and the latter is rarely complete; a certain amount of residual magnetism remains, hysteresis being strictly due to this retention of energy (Sprague). Hysteresis shows itself in heat, but must not be confounded with Foucault or eddy currents. The latter are corrected by subdividing the metal, but the former depends upon the quality of the metal, and increases with its length.

Moreover, a coil with a closed magnetic circuit requires an independent contact breaker.

In most of the alternating currents used in lighting their rapidity of alternation is but one hundred and twenty-five periods per second. As in the simple electro-magnet, the proportions of diameter and length of the primary coil and core will determine its rapidity of action. A short fat coil and core will act much quicker than a long thin one. But on a short fat coil the outside turns would be too far removed from the intensest part of the primary field. A good proportion of core length is given in the following table:

Spark Length of Coil.	Iron Core.
$\frac{1}{4}$	$4'' \times \frac{1}{2}''$
$\frac{1}{2}$	$5'' \times \frac{10}{16}''$
1	$7'' \times \frac{3}{4}''$
2	$9'' \times 1''$
3	$12'' \times 1\frac{1}{8}''$
6	$19'' \times 1\frac{1}{4}''$

The primary coil P consists of two or not more than three layers of insulated copper wire of large diameter, being required to carry a heavy current in a 2-inch spark coil, probably from 8 to 10 amperes. In designing the primary coil a great advantage arises from using comparatively few turns but of large wire. Each turn of wire in the primary has a choking effect upon its neighbor by what is termed self-induction.

As the primary coil and core may be considered as an electro-magnet, it may not be out of place to notice the rule governing such. Magnetization of an iron core is mainly dependent upon the ampere turns of the coil surrounding it—that is, one ampere carried around the core for one hundred turns (100 ampere-turns) would equal in effect ten amperes flowing through ten turns. Practically speaking, there would be certain variations to the rule, for one difficulty would arise in that the smaller wire used in conveying the smaller current would fit more compactly and allow more turns to be nearer the core, the active effect of the turns always decreasing with their distance from the core. And although a large current and few turns would not have so much self-induction, there would be trouble at the contact breaker, owing to the large current it would have to control.

The most suitable sizes of wire for the primary coil are: No. 16 B. & S. for coils up to 1 inch spark; No. 14 B. & S. up to 4 inches of spark, and No. 12 B. & S. for a 6-inch spark coil. The coil should be, say, one-twelfth of the core length shorter than the core.

I is the insulating tube between the primary coil and the secondary coil S. Here great precaution is necessary to prevent any liability of short circuiting or breaking through of sparks from the secondary coil. This danger cannot be underestimated, and the tube should be either of glass or hard rubber, free from flaws, varying in thickness with the dimensions of the coil. It should extend at least one-tenth of the total length of the primary coil beyond it at each end. The end of this tube can be turned down so as to allow of the hard rubber reel ends being slipped on and held in position by outside hard rubber rings (Fig. 2).

The secondary coil consists of many turns of fine insulated copper wire separated from the primary coil by the insu-

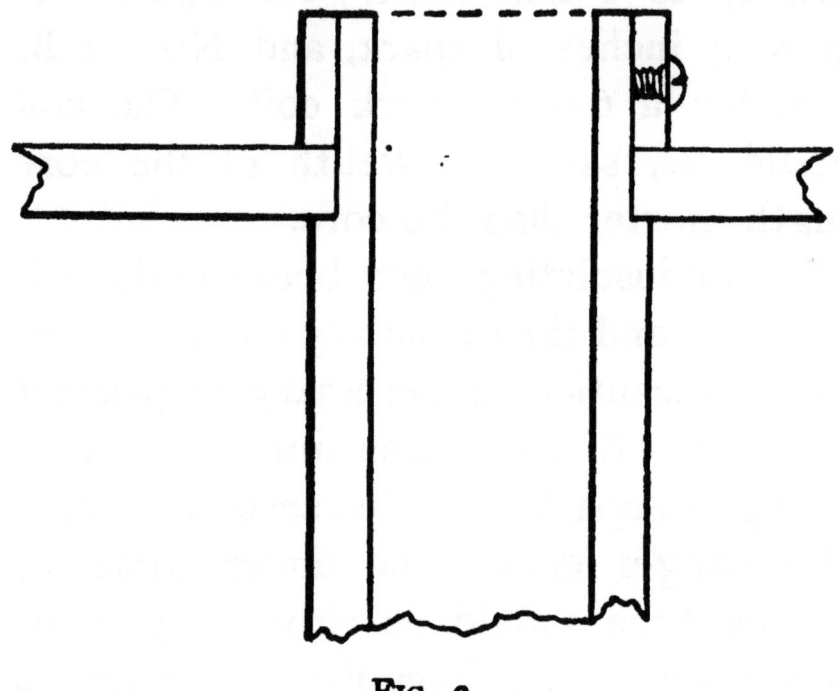

FIG. 2.

lating tube and a liberal amount of insulating compound at each end. In coils giving under 1 inch of spark this coil may be wound in two or more sections.

The usual manner of constructing these

Coil Construction.

sections is to divide up the space on the insulating tube by means of hard rubber rings placed at equal distances apart, in number according to the number of sections desired (Fig. 3). The space between each set of rings, or between the coil end

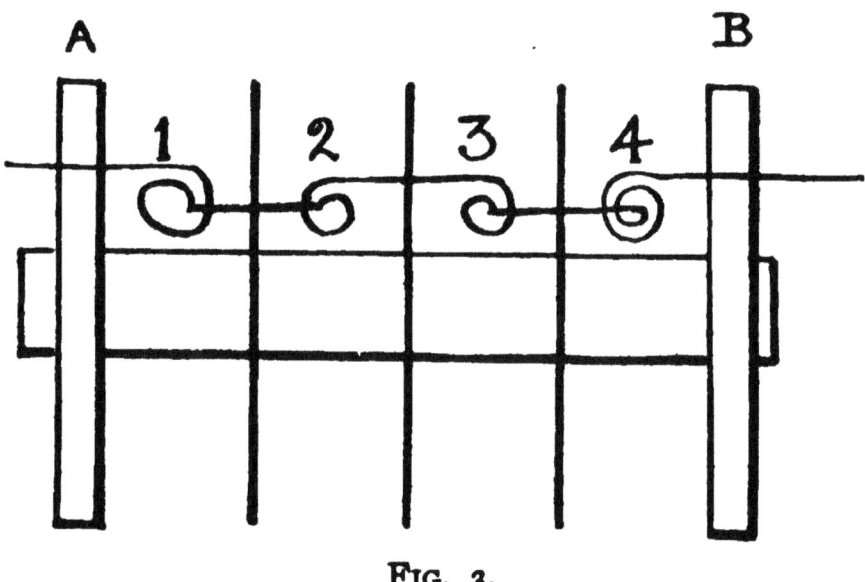

FIG. 3.

and a ring, is wound with the wire selected, the filled sections constituting a number of complete coils, which are finally connected in series. The sectional method of winding prevents the liability of the spark jumping through a short circuit, but heightens its tendency to pass into the primary coil at the ends, where it must be therefore specially insulated from it.

In winding these sections there is a

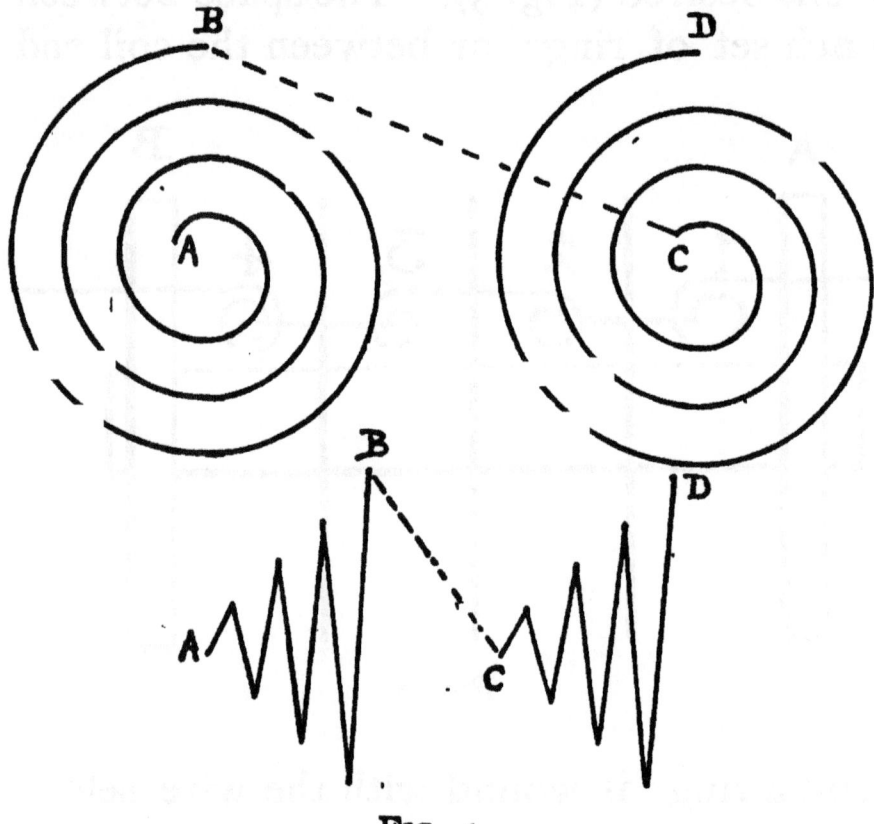

FIG. 4.

method now generally adopted which has many good points, although at first it may seem complicated. . The old way of filling two sections was to wind both in the same direction as full as desired, then join the outside end of the left-hand coil to the inside end of the right-hand coil. This necessitated bringing the outside end down between two disks, or in a vertical

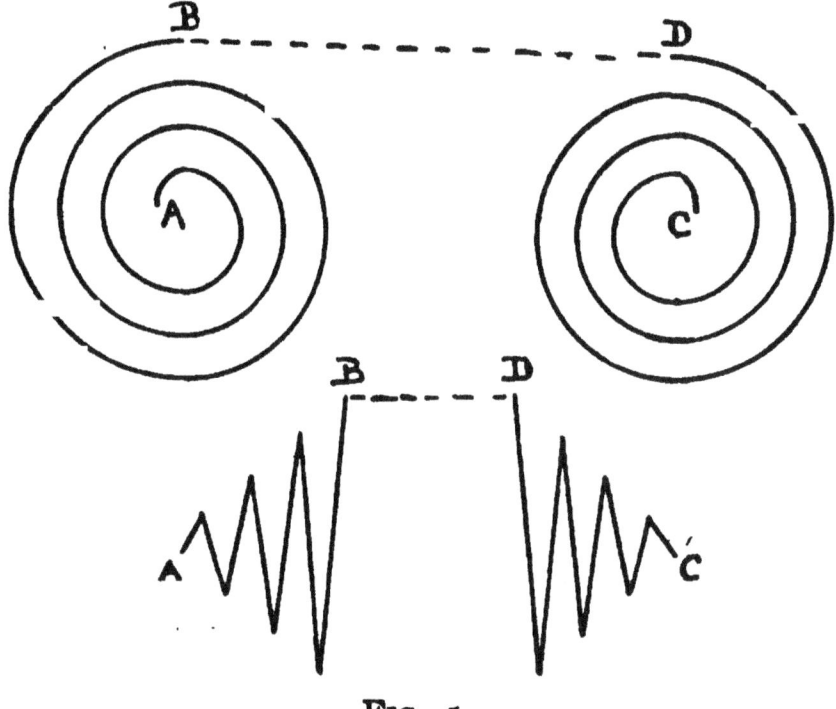

Fig. 5.

hole in the sectional divider, and thereby rendered it liable to spark through into its own coil. This is shown in Fig. 4, *A* and *C* inside ends, *B* and *D* outside ends, the disk being between *B* and *C*.

Reference to Fig. 3 shows the new method, and Fig. 5 shows an enlarged diagram of sections 2 and 3 of Fig. 3.

Sections 1 and 3, Fig. 3, are filled with as many turns as desired; the spool is then turned end for end, and sections 2 and 4 are wound, being thus in the opposite direction of winding to sections 1 and 3.

The inside ends of 1 and 2 and 3 and 4 are soldered together, and the outside ends of 2 and 3 are also soldered together.

The outside ends of 1 and 4 serve as terminals for the coil.

This method of connection leaves all the turns so joined that the current circulates in the same direction through them all, as will be seen by an examination of the enlarged diagram, Fig. 5.

Sprague, in his " Electricity : Its Theory, Sources, and Application," recommends that the turns of wire in the secondary coil shall gradually increase in number until the middle of the spool is

Coil Construction. 15

reached, and then decrease to the spool end, in order that the greatest number of turns be in the strongest part of the magnetic field (see Fig. 6). *D D D* are section dividers, *S* secondary windings, *P* primary coil. The selection of the size of wire to

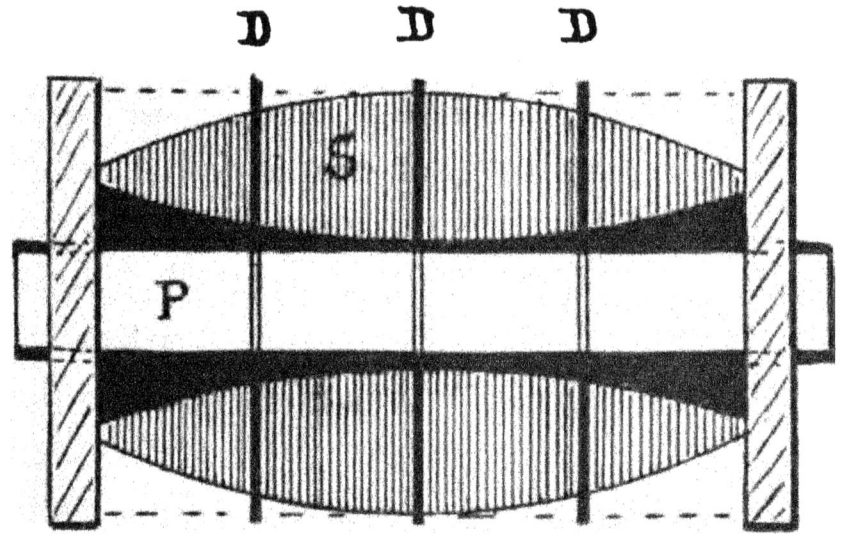

FIG. 6.

be used depends on the requirements as to the spark. If a short thick spark be desired, use a thick wire, say No. 34 B. & S.; if a long thin one, use No. 36 to No. 40 B. & S.

Although it is impossible to lay down

rules for determining the exact amount of wire to be used to obtain a certain sized spark, yet a fair average is to allow 1¼ pounds No. 36 B. & S. per inch spark for small coils and slightly less for large ones.

The most satisfactory and perhaps the easiest way for large coils is to wind the secondary in separate coils, made in a manner similar to that employed in winding coils for the Thompson reflecting galvanometer. This method, first described by Mr. F. C. Alsop in his treatise on "Induction Coils," is somewhat as follows:

A special piece of apparatus (Figs. 7 and 8) is necessary, but presents no great difficulty in manufacture. A metal disk, D, one-sixth of an inch thick and 7 inches in diameter, is mounted on the shaft S. A second disk is provided with a collar and set screw, A, in order that it may be adjusted on the shaft at any desired distance from the stationary one. When the diameter

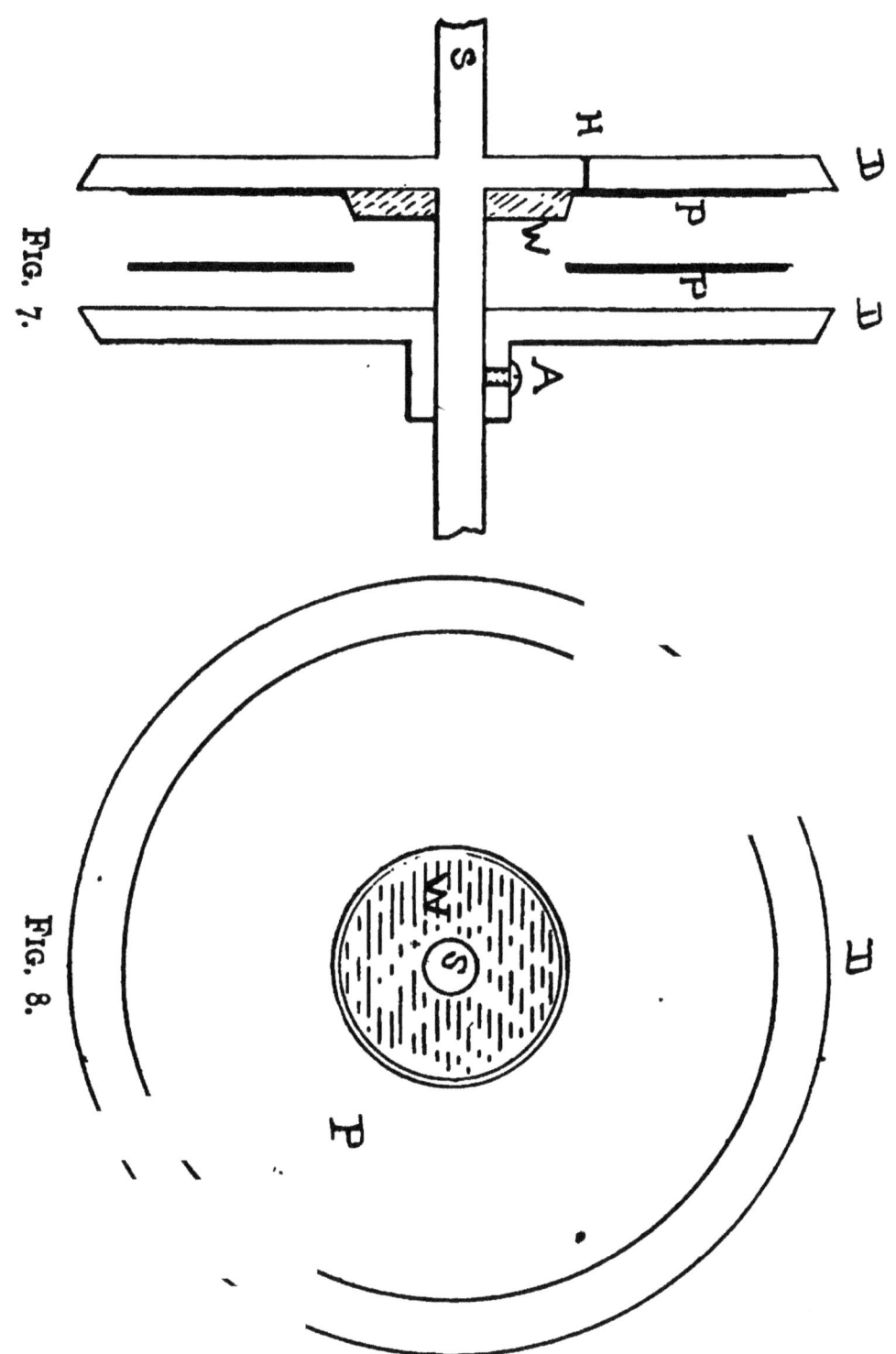

Fig. 7.

Fig. 8.

of the coil to be wound has been decided upon, a wooden collar, *W*, with a bevelled surface is slipped on the shaft, it corresponding in diameter with the desired diameter of the hole through the centre of the secondary coil. As these coils are going to be made as flat rings and slipped on over the insulating tube, a remark here becomes necessary on this diameter. Reference to Fig. 9 will show that it is intended that the coils near the reel ends shall fit very loosely on the tube *T* (Fig. 1)—in fact, that there shall be a clearance of possibly one-half inch in the extreme end, diminishing gradually to a fifteenth of an inch in the centre coils. Therefore it becomes necessary to provide a number of wooden rings equal to the desired diameter of the central hole in the coil. The thickness of

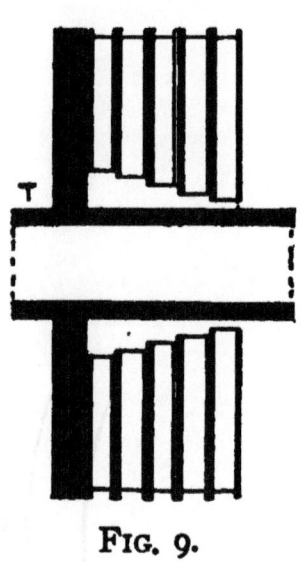

FIG. 9.

the wood determining the width of the individual coil depends on the selection of the operator; but the rule may be laid down that the narrower the coils the better the insulation of the complete coil will be on completion.

One-sixteenth of an inch is a very fair average, and has been generally adopted by the writer..

A quantity of paper rings are now cut out of stout writing paper which has been soaked in melted paraffin. If a block or pad of letter paper be soaked in paraffin and allowed to become cold under pressure, the ring may be scratched on the surface of it and the block cut through on a jig saw. The central apertures of course will vary in size with their position on the tube T (Fig. 9).

The coil winder is now either mounted in a lathe or fixed in a hand magnet winder in such manner that it can be steadily and rapidly rotated. The wire to be wound comes on spools, which can be so threaded on a piece of metal rod that they turn readily. A metal dish containing melted paraffin is provided with a round rod, preferably of glass, fixed under the paraffin surface, so that it can rotate freely when the wire passes under it through the paraffin. Two paper rings are slipped on the winder that they may form, as it were, reel ends for the coil, and if the metal disks have been warmed it is an easy matter to lay them flat.

The end of the wire is then passed through the paraffin under the glass rod and through the hole H in the metal disk for a distance of, say, 6 inches, and held to the disk outside with a dab of paraffin or beeswax. Then the winder is rotated and the space between the paper disks is filled with wire. The paraffin, being hot, will adhere to the wire, and cooling as the wire lays down on the winder, hold the turns together and at the same time insulate them from each other. It will not be possible to lay the wire in even layers, as would be necessary in winding a wider coil, but the spaces can be filled up, using ordinary care that no radical irregularity occurs—that is, that only adjacent layers are likely to commingle.

When the space is filled up to the level of the paper disks and the paraffin is hard, loosen the set screw, and removing the outside disk, the coil can be slipped off, or a slight warming will loosen it. Any number of these coils can be made, and there are the advantages in this mode of construction that a bad coil will not spoil the whole secondary, and that the wire can be obtained in comparatively small quantities.

As each coil will not be of very high resistance, the continuity of the wire can be readily tested by means of a few cells of battery, connecting one end of the coil to one pole of the battery, and the other pole of the battery and coil end touched to the tongue. If a burning sensation is experienced, the connection is not broken.

Where possible the coils should be measured as to their resistance on a Wheatstone bridge.

When the requisite number of coils has been prepared, they are assembled in the following manner (Fig. 9): The coils, having their aperture diameter graded, are placed in order, and starting with the one having the largest hole, it is slipped over the primary protection tube T, one end being brought out through a hole in the reel end drilled vertically or between the reel end and the coil. A couple of paper rings are then slipped on the tube, and another coil placed over them, having its ends connected as in Fig. 3. This process is continued until all the coils are in place. The annular space between the coils and the tube T (Fig. 9) is filled in with melted paraffin and the coils gently pressed together, so as to form a compact mass, paraffin being poured over the outside of the whole combination. Before winding any wire used in this work it must be perfectly dry, which end can be accomplished by subjecting the whole spool to a short period of baking in a moderately warm oven.

The accompanying table gives the length of No. 36 silk-covered wire that will fill a linear space equal to one thickness of the wire in different-sized rings. This size wire wound tight will give 125 turns per linear inch. Therefore on a ring having a middle aperture of 1½ inches and an outside diameter of 4 inches, there will be 156 turns, or a total length of 1347 inches. This is obtained as follows: 1½ inches × 3.1416 = 4.7124 (or 4.712); 4 inches × 3.1416 = 12.5664 (or 12.56); $\dfrac{4.712 + 12.56}{2}$ = mean circumference—viz., 8.635 inches.

This mean × number of turns in thickness of ring between the two circumferences—viz., 156 = 1347 inches.

To obtain the length of wire necessary for a ring occupying more than the space

Coil Construction.

No. 36 Silk-Covered Wire. 125 Turns per Linear Inch. 13,306 Feet per Pound.	1¾" Aperture Diameter, 4.712" Aperture Circumference.			2" Aperture Diameter, 6.283" Aperture Circumference.			2½" Aperture Diameter, 7.854" Aperture Circumference.		
	4"	5"	6"	4"	5"	6"	5"	6"	7"
Outside diameter............									
Outside circumference.......	12.56	15.70	18.84	12.56	15.70	18.84	15.70	18.84	21.99
Mean circumference.........	8.635	10.20	11.78	9.42	10.99	12.56	11.78	13.35	14.92
Turns between circumferences........	156	219	282	125	188	250	156	219	282
Distance between aperture and outside, in inches.........	1.25	1.75	2.25	1	1.50	2.	1.25	1.75	2.25
Length of wire, in inches.....	1347	2234	2650	1178	2066	3140	1838	2924	4207

of one turn on the primary insulating tube, multiply the length before obtained by the number of turns in the space it occupies. Thus a flat ring one-tenth of an inch thick would equal 1347 inches × 12.5.

This rule is necessarily only approximate, owing to the way the wires bed on each other from their cylindrical section. In actual practice, when the wire is run through the paraffin bath not more than 50 per cent of the calculated wire will occupy the space. And the thickness of the paper rings must also be added when figuring the total length of the coil. In the iron-clad transformers or induction coils of highest efficiency used in the alternating current electric light system, the rule for determining the windings of the coils is based on the ratio of the turns of wire in the primary to the turns in the secondary, the electromotive force in the primary, and the lines of force cut by the windings.

The secondary ends can be attached to binding posts mounted on the reel ends.

Unless these reel ends be very high and clear the outside of the coil considerably, it is better to mount the binding posts on the top of the hard rubber pillars. A neat plan is to mount on the top of the coil a hard rubber plate reaching from reel end to reel end, and place the binding posts on that.

A discharger consists of two sliding metal rods with insulated handles passing through pillars attached to the secondary coil. The inside ends of these rods is provided with screw threads for the ready attachment of the balls, points, etc., which are to be used. The substance to be acted upon is laid on a rubber or glass table midway between the rod pillars and slightly below the level of the rods.

By hinging the rod pillars, or using a ball and socket joint, the discharger can be inclined so as to be better brought near the substance on the table.

The next important part of the coil is the contact breaker.

The armature R is a piece of soft iron carried at the end of a stiff spring, in about the middle of which, at B, is riveted a small platinum disk or stud. The adjusting screw A has its point also furnished with a piece of platinum, which is intended to touch the platinum on the spring when the latter is in its normal position. The core C of the coil serves as an electro-magnet. When the current flows from the battery (represented by the figure at L) through the primary coil and armature spring to the adjusting screw, it causes the armature to be drawn to the magnetized core, but thereby draws the platinum disk away from the adjusting screw. In so doing it breaks the circuit, the magnet loses its power, and the elasticity of the spring reasserting itself, carries the armature back, thereby reclosing the circuit. This is repeated many times in a second, the result being a continual vibration of the spring, and a consequent interruption to the current.

The condenser or Leyden jar *J*, connected as in the diagram to the base of the vibrating spring at *K* and to the adjusting screw wire *M*, is constructed as follows: On a sheet of insulated paper is laid a smaller sheet of tinfoil, one edge of which projects an inch or so over one end of the paper. Another sheet of paper covering this carries a second sheet of tinfoil, one end of which projects as in the first sheet, but at the opposite end of the paper. Tinfoil and paper sheets are laid in this manner alternately until a sufficient number is attained. The projecting ends are then clamped together and the whole pile immersed in melted paraffin, as will be described in a subsequent chapter. Wires are affixed to these clamped ends which serve to connect the condenser with the contact breaker. The conventional sign for a condenser is that used at *J*, showing the two series of plates, the insulation or dielectric, as it is called, being understood.

The size of condenser to use with different-sized coils varies according to the winding of the primary and the battery used. A primary coil of few turns would not necessitate as large a condenser as one of a large number of turns. At the same time, a condenser may be made of too great a capacity, and thereby weaken the action of the coil.

The base upon which the coil and its parts are mounted may be of dried polished wood. But where the coil is designed to give large sparks—over 2 inches—it is an advantage to use hard rubber one quarter of an inch and upward in thickness. Glass, were it not for the difficulty of drilling it and its brittleness, would be a desirable material for a coil base in a dry atmosphere. Hard red or black fibre coated with shellac varnish is also serviceable, and, moreover, is extremely easy to work. Slate must never be used; there is too much liability of iron veins being found in it, which in such high tension experiments as will be described would seriously impair the usefulness of the apparatus. The material selected for the base must be one that will not absorb moisture. A paraffined surface collects moisture up to a certain point in isolated drops, whereas a glass and even a hard rubber surface condenses the moisture as a film, which latter is extremely undesirable. But unfortunately the fact that a paraffined surface does not present a pleasing appearance would probably result in its rejection. And lastly, by mounting the coil on hard rubber blocks, or extending the reel ends to raise the coil body, a high insulation can be obtained at the sacrifice perhaps of appearance or height. From the care taken to insulate the secondary coil, it may be considered a superfluous precaution to so carefully select a base, but practical work with the instrument at some important crisis will demonstrate the necessity of extreme care in the smallest details relating to insulation. It may be well to note here that hard rubber is acted upon by ozone, and is thereby impaired as an insulator.

The base forms the top of a flat box in which the condenser lies; but there are a few points worth considering right here. As the connections of the coil will probably be under the base, a sufficient space must intervene between the base and the top of the condenser. It is a good plan to lay the condenser at least one half inch below the top of this box, and fill up to, say, one eighth of an inch with melted paraffin, leaving the condenser wires projecting for attachment. The connections of the primary coil and contact breaker should by all means be soldered, not simply wires held under screw nuts. And, moreover, all wires under the base should be so run that they do not cross one an-

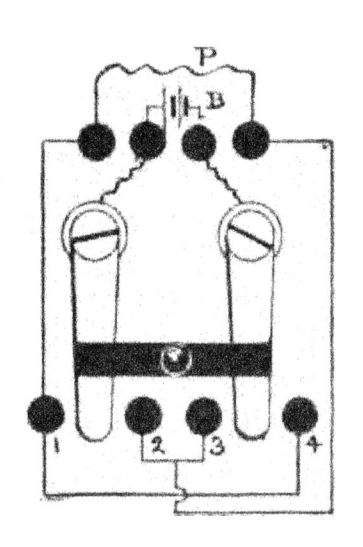

FIG. 10.

other, which precaution only requires a little planning. Then, when the connections are all made and the base laid on top of the box, it can be pressed down if the paraffin be warm, so that the screw heads and wires mark out their own channels and cavities in which to lie.

A commutator or pole-changing switch is often added to change the polarity of the battery current. The diagram of connection is shown in Fig. 10. When the levers are as in the figure, the circuit is broken and no current flows through the coil.

Coil Construction.

Gauge, Browne & Sharpe.	Diameter.	Feet per lb.	COPPER. Ohms per 1,000 ft.	GERMAN SILVER. ONLY APPROXIMATE. Ohms per 1,000 ft.
8	.1285	20	.62881	11.77
9	.1144	25	.79281	11.83
10	.1019	32	1	18.72
11	.09074	40	1.2607	25.59
12	.08081	51	1.5898	29.75
13	.07196	64	1.995	37.51
14	.06408	81	2.504	47.30
15	.05707	102	3.172	59.65
16	.05082	129	4.001	75.22
17	.04525	162	5.04	94.84
18	.0403	204	6.36	119.61
19	.03539	264	8.25	155.10
20	.03196	325	10.12	190.18
21	.02846	409	12.76	239.81
22	.02535	517	16.25	302.38
23	.02257	660	20.30	381.33
24	.0201	823	25.60	480.83
25	.0179	1039	32.20	606.31
26	.01594	1310	40.70	764.59
27	.01419	1650	51.30	964.13
28	.01264	2082	64.80	1215.76
29	.01126	2623	81.60	1533.06
30	.01002	3311	103	1933.03
31	.00893	4165	130	2437.23
32	.00795	5263	164	3073.77
33	.00708	6636	206	3875.61
34	.0063	8381	260	4888.49
35	.00561	10560	328	6163.97
36	.005	13306	414	7770.81

CHAPTER II.

CONTACT BREAKERS.

THE simple form of contact breaker already described is useful up to a certain point, but it has disadvantages. Its rate of vibration is only variable through narrow limits, and it is not suitable for very heavy currents. But as it stands it has done long service, and will be used probably wherever the requirements from it are not exacting. The most desirable form of this

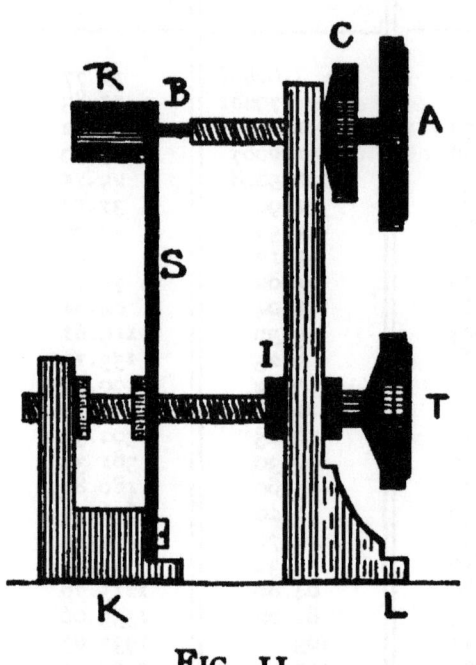

FIG. 11.

simple spring break is shown in Fig. 11. *R* is the soft iron armature; *S*, the spring; *C*, check-nut which holds the adjusting screw *A* from becoming loose; *T*, a second adjusting screw used to tighten the spring and so raise its rate of vibration; *K* is the base to which one wire of the coil is attached; *L*, base of adjusting device to which battery wire runs at *I*. Where tightening screw *T* passes through the pillar of the adjusting screw, the hole therein is bushed with rubber to prevent accidental contact. Both *A* and *T* are provided with insulating heads of rubber or ivory. At *B* are the platinum contacts, which should be fully $\frac{1}{8}$ inch in diameter.

One serious defect in the action of the simple spring vibrator (Fig. 12) is the tendency of the spring to vibrate, as it were, sinusoidally. This causes an irregularity in the rate of the vibrations, which affects the discharge of the coil very considerably. By far the better plan is to use a very short thick spring riveted to an arm carrying the armature at its end (Fig. 13). *R* is the armature, *S* the piece of spring, and *K* the point of attachment to the base. The actual width of the portion of the spring which vibrates—the hinge

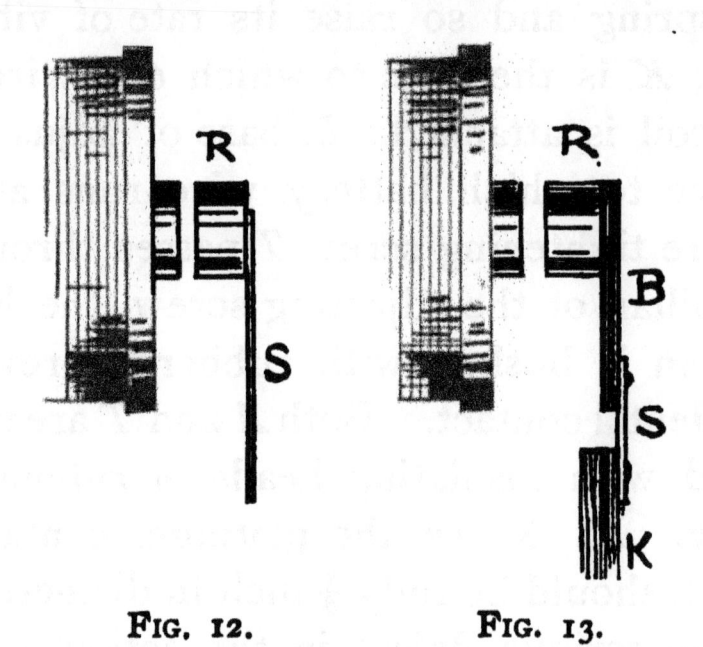

FIG. 12.   FIG. 13.

portion, it might be called—should not be over ⅛ inch.

The rate of motion is high; but an erroneous notion has been taken of its performance by many persons in the knowledge of the writer. The rate of vibration is *not* wholly dependent on the size, or, rather, smallness of its spring; the arm and armature considerably alter this, although they are not pliable, by reason of their mass and the momentum consequent on their mass.

A word here on the size of the armature. It should be somewhat larger than the face of the electro-magnet core, and should be thick—that is, in a circular form—say one half its diameter. Of course this does not apply to the steel lever armature before mentioned. It is impossible to lay down arbitrary rules where the conditions are not determined, but a very small amount of experimenting will demonstrate the correct lines on which to build.

When in action, all rapid rheotomes give out a definite musical note whereby the rate of vibration can be determined. Reference to any work on acoustics will show a table of the number of vibrations necessary to produce any stated musical note. The foregoing style of rheotome

38 Contact Breakers.

forms the basis of very nearly all those which are in use. The shorter and stouter a spring the more rapidly will it vibrate, and *vice-versa*. Carrying out this rule, we can manufacture an instrument which

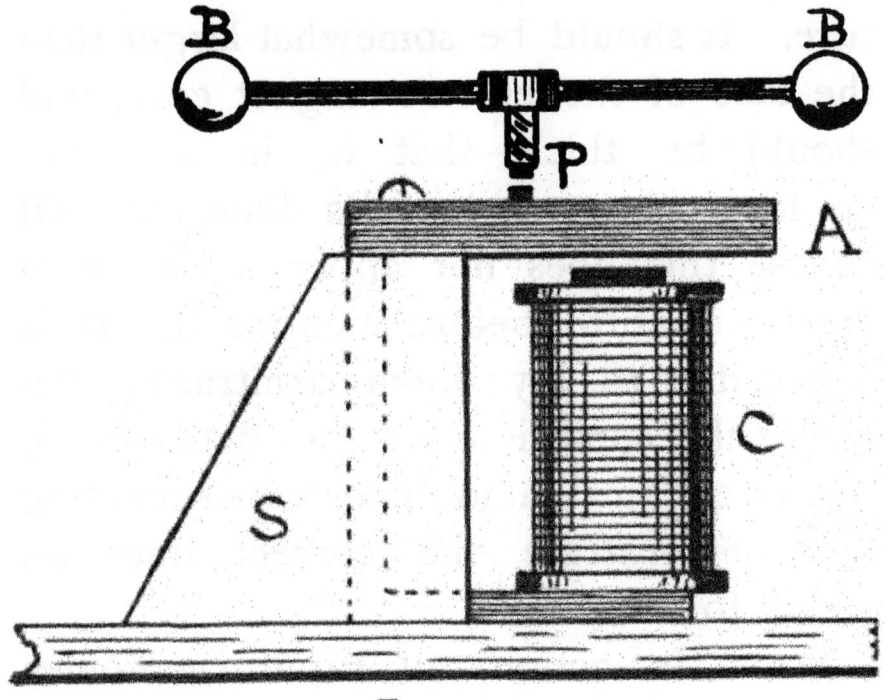

FIG. 14.

will give as high as 2500 vibrations per second (Fig. 14).

The armature A is a piece of flat hard steel bar $\frac{1}{4} \times \frac{1}{2}$ inch, held rigidly on the metal support S and just clearing the upper surfaces of the magnet cores *C*. The adjusting screw *P* should be provided with an arm, *B B*, whereby the rotation of it can be delicately varied. This screw must also be firmly held or the high speed of the armature will jar it loose. A check-nut on each side of the frame carrying it should be provided in every case. The necessary platinum contact can be hammered into a hole drilled before the armature is hardened. The proper place for this contact is about one fourth of the total length of the armature from its support, although in the simple contact breaker it can be placed at the distance of one third if desired. The reason is that the concussion of the adjusting screw dampens the free vibration, and the amplitude thereof is lessened in addition to the counter vibrations of the screw disturbing the regular vibrationary series.

Owing to the fact that the amplitude of the armature vibration is so small, a very delicate adjustment is necessary. The adjusting screw can be placed nearer the free end, but for the reasons given it is not to be desired. The metal bridge should be a solid casting. and the armature clamped by more than one screw.

The mercury vibrator, which is applied to almost every large coil, is as follows:

A pivoted arm carries on one end a soft iron armature, which is attracted by the coil core. The other end is provided with a platinum point adjustable by a set screw. This platinum point dips into a mercury cup—a glass cup containing mercury, with a thin layer of spirits of turpentine. The object of the spirits of turpentine, which is a non-conductor, is to help choke off the spark which would ensue whenever the platinum point was raised from the mercury.

A form of contact breaker which will admit of great variation of speed, and which is adapted to carry large currents, is the wheel-break, constructed in the following manner:

Contact Breakers.

A brass or copper disk 3 inches or more in diameter and upward of $\frac{1}{8}$ inch thick has its periphery divided by a number of saw cuts, which divisions are often filled in with plugs of hard rubber or fibre. This disk is mounted on a shaft, which latter is either the shaft of an electro-motor, or is provided with a pulley by which it can be rapidly rotated. A strip of spring copper on each side of the disk presses upon the toothed surface, one strip being connected to the coil and the other to the battery or other current source. It will now be seen that when the disk rotates the slits or pieces of hard rubber cause the break in the circuit through the brushes or copper strips, the rapidity of the breaks depending upon the rate of rotation of the disk, and the number of slits in the wheel.

The slits or rubber pieces should be one-half the width of the intervening brass, but must be at least one sixteenth of an inch in width, especially where a high voltage is used in the primary coil.

The shaft of the machine may serve as one point of connection in place of one of the copper brushes; but in this event either a wide journal must be used, or else some conducting substance, as plumbago, replace the lubricating oil in the bearings.

POLE CHANGING BREAKER.

Fig. 15 shows a diagram of a pole changing contact breaker which will allow of rapid alternations of current. It is operated by an electric motor by preference, although any motive power can be applied to it.

$W a\ W b$ are two brass wheels, the peripheries of which are broken by the insertion of insulating blocks $I\ I$, shown black in the sketch. $S\ S$ are the shafts on which the wheels are mounted, the two wheels being necessarily insulated from each other. 1, 2, 3, 4 are four brushes of copper pressing on the rim of the wheel and leading in the current from the battery B.

FIG. 15.

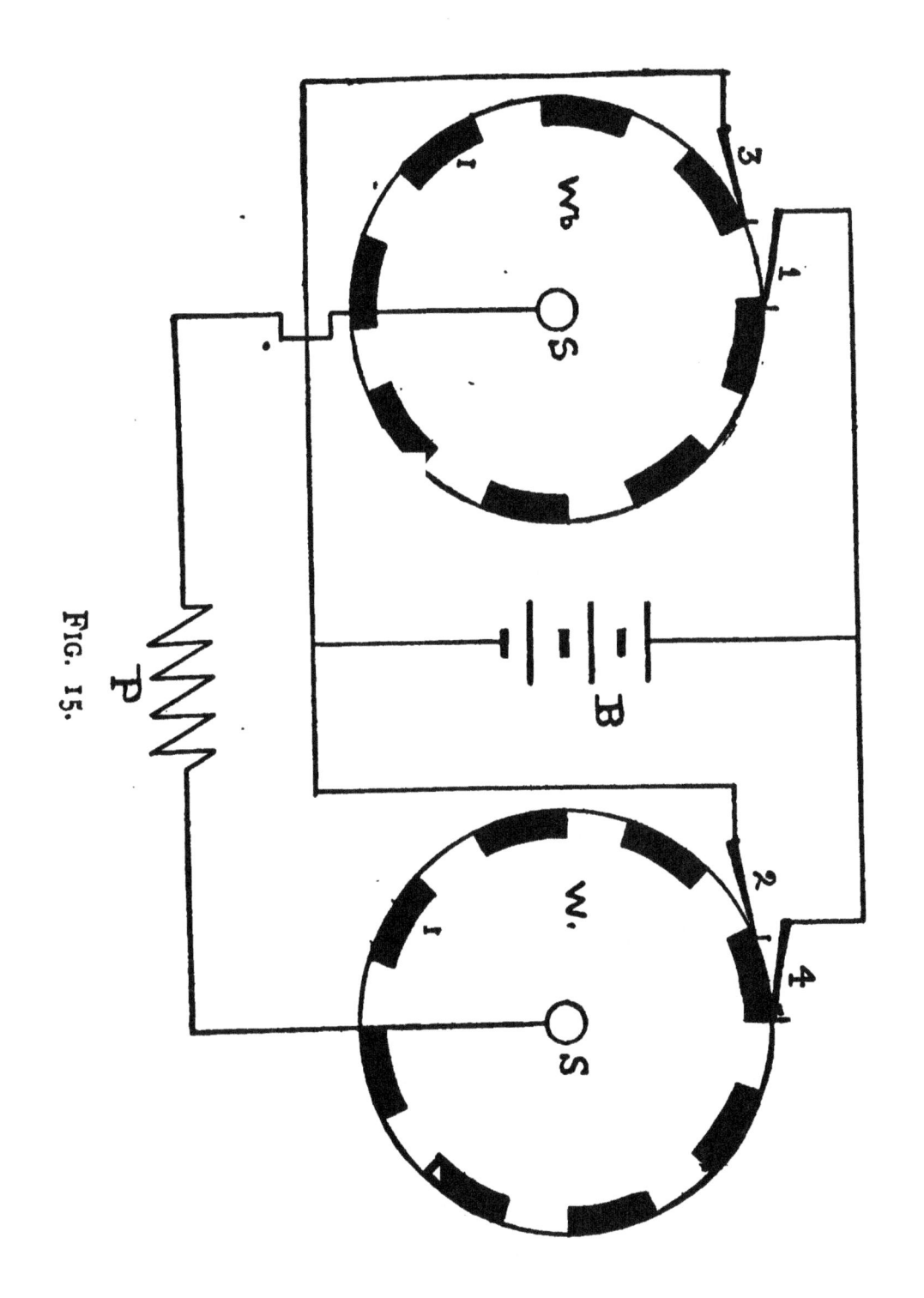

The primary coil is attached to the brass body of the wheel or to the shafts. When the wheel is in the position shown, the coil and battery are on an open circuit; but on the wheel commencing to revolve, the brushes 1 and 2 bear on the brass, and the current flows from the positive pole of the battery to 2 through the wheel Wa to the coil P, up through wheel Wb and out at 1 back to the battery. The next position of the brushes 1 and 2 will be on the insulations, and 3 and 4 will come into action. Then the positive current will reach Wb by means of brush 3, and after traversing the primary coil and wheel Wa, emerge at 4 to the battery, thus reversing the current through P as many times as there are sets of segments, which latter can be multiplied according to requirements. The main point to be considered after that of good connections is that the brushes 1 and 3 and 2 and 4 do not at any time touch any part of the brass wheel at the same time, as this would short circuit the battery. This is avoided by making the insulating space longer than the brass surface, and adjusting the brushes as in the sketch, that each pair of them is a fraction further apart than the length of the brass tooth.

Accordingly, a wheel may be constructed with many segments and rotated at a high speed and rapid reversals of current produced, the uses of which are manifold.

As will be described in the notes on the Tesla effects, an electro-magnet, the poles of which are brought near the sparking point of the contact breaker, will help wipe out the spark, and so assist the suddenness of the break.

An extremely successful expedient in operating contact breakers is to employ a high-pressure air blast directed point blank against the contact point. The effect of this air blast when the contact is made is of course null, but on the platinum surfaces becoming separated, the high air pressure produced forms a path of extremely high resistance, and tends to blow off the spark as soon as it is generated. The stream of air should issue from an insulated nozzle of glass or rubber, and should not contain moisture.

CHAPTER III.

INSULATIONS AND CEMENTS.

IN selecting an insulating compound for apparatus designed to be under the influence of high tension currents, a glance at some of the peculiarities of such currents will not be out of place. Mineral oil is used in many of the converters employed to transform the high voltage currents on the mains of the alternating electric-light systems to the comparatively low voltage used at the points of consumption. Professor Elihu Thomson, in a series of experiments, noticed some interesting facts in the sparking distances of high potentials in oils.

He found that discharges of low frequencies, as 125 alternations per second, were capable of puncturing mineral oils at one third to one half the thickness of an air layer sufficient to just resist punctures by the same discharge; but with frequencies of 50,000 to 100,000 per second, an oil thickness of one thirtieth to one sixtieth was a sufficient barrier.

At a frequency of 125 per second, a half-inch spark in the air penetrated one third to one fourth inch of oil; but at frequencies of 50,000 to 100,000 per second, a layer of oil one fourth of an inch successfully resisted the passage of a spark which freely passed through 8 inches of air.

The effect of drying an oil improved its insulating qualities. (Tesla uses boiled-out linseed-oil.)

He also noted that pointed electrodes could be brought nearer together under oil than balls without allowing a discharge. Flat plates allowed of still greater sparking distances. Tesla notes that oil through which sparks have passed must be discarded, probably owing to particles of carbon being formed.

Paraffin wax has a higher resistance than oil, providing it has not been heated over 135° C. It will stand alternate heating up to 100° C. and cooling, being of lower resistance when hot than when cold. But a serious permanent deterioration takes place when it has been heated over 100° C.; its color, from the normal pure white, changes to a yellowish tint when its insulation is impaired. Paraffin also undergoes a deterioration when heated for a long time even at 100° C., and should never be used for fine work when it is at all yellow. It is always best to melt it in a hot-water bath, not permitting, however, any steam or moisture to come near it. In this climate (United States) it is not so necessary to mix in any tallow to obviate brittleness, the average temperature of most workshops being sufficiently high to keep it from becoming brittle.

Resin oils do not suffer permanent injury from being heated, as does paraffin, but their insulating properties diminish much more rapidly on becoming even warm, the initial resistance of resin oils being lower than that of paraffin.

Paraffin has a fault—its tendency to absorb a slight degree of moisture. It has been found in telephone and telegraph cables saturated with paraffin that this is a very important cause of their deterioration. In Ruhmkorff coils, however, which are intended for operation in enclosed places free from damp atmospheres, the absorption of moisture would be probably reduced to its minimum.

There is one substance which, were it not for its cost, would be far preferable to paraffin for coil work, and that is beeswax. Its cost, however, is generally five times that of paraffin, even when purchased in quantities. It never becomes brittle enough to be damaged in careful handling, its melting point is low, and it does not absorb moisture. But it must be unquestionably pure and clear.

In foreign practice a variety of resinous

mixtures are used to insulate the turns of the wire in Ruhmkorff coils.

Equal parts of resin and beeswax used hot, paraffin, resin and tallow, and shellac and resin are employed.

Shellac—that is, the yellow lac—is much used as a varnish for electrical instruments, being dissolved in alcohol to saturation. For dynamo armatures and similar apparatus the shellac varnish is of great service, and many good compounds of shellac, such as insullac and armalac, have been prepared for ready use. But (excluding beeswax) for our purposes paraffin stands pre-eminently at the head of the list.

In using shellac varnish, in high tension work more particularly, care must be taken that the moisture has entirely evaporated. Although a piece of shellacked apparatus may appear perfectly dry, yet when the current is allowed to flow unlooked-for results may appear—it takes hours in a dry atmosphere for shellac varnish to dry. Baking the apparatus in a warm oven is a necessary expedient whenever feasible, care being taken not to burn or decompose the shellac. The proportions most generally used are 1 ounce shellac to 5 ounces alcohol. Stand the vessel containing the mixture in a warm place, and shake it frequently; filtration improves the varnish somewhat.

A ready and efficient varnish for silk is prepared by mixing 6 ounces of boiled linseed-oil and 2 ounces of rectified spirits of turpentine. For paper, 1 part of Canada balsam and 2 parts of spirits of turpentine dissolved in a warm place and filtered before being used. A good insulating cement for Leyden jars and insulating stands is prepared from sulphur, 100 parts; tallow, 2 parts, and resin, 2 parts, melted together until of the consistence of syrup, and sufficient powdered glass added to make a paste. To be heated when applied, this will resist most acids. The resin and beeswax compound is handy when making experimental mercurial air pumps of glass tubes, as it has a fair tenacity, is not too brittle, and is easily used.

CHAPTER IV.

CONDENSERS.

A CONDENSER is an apparatus whereby a charge of electrical energy may be temporarily stored, the amount of energy it will hold determining its "capacity." The capacity of a condenser is measured in micro-farads, the commercial unit representing one millionth of a farad. A farad equals the capacity of a body raised to the potential of one volt by a charge of one ampere for one second at one volt—*i.e.* = one coulomb.

The measurement of the capacity of a condenser is accomplished by the use of a ballistic galvanometer. The latter instrument has a bell-shaped magnet suspended in a coil of fine wire. When a momentary current is passed through this coil the magnet hardly commences to rotate until the current has practically ceased. A beam of light is reflected from a mirror fixed to the magnet on to a scale. The degree of deflection is compared with that obtained by the discharge of a condenser of known capacity, and the capacity of the condenser being measured is deduced by a simple rule. The farad, which is the unit of capacity requiring a condenser of an immense size, is replaced by a commercial unit, the micro-farad—that is, one millionth of a farad.

The original form of the condenser was the Leyden jar, which owes its name from the town of Leyden in Europe.

The Leyden jar is made as follows (Fig. 16): A clean uncracked glass jar with a wide mouth is coated on the inside and outside with tinfoil; sometimes loose tinfoil is filled inside, the tinfoil, however, not reaching more than two thirds of the jar's length from the bottom. A cork is fitted, and through the middle of it a wire is passed touching the inside coating of tinfoil and terminating in a metal sphere outside. A simple Leyden jar can be made in a few moments by half filling a glass bottle with water and wetting the lower half of the outside; a wire run through the cork into the water finishes the job. But this is at least only a makeshift, although a fair amount of current has been collected from a leather engine belt in motion in one thus made.

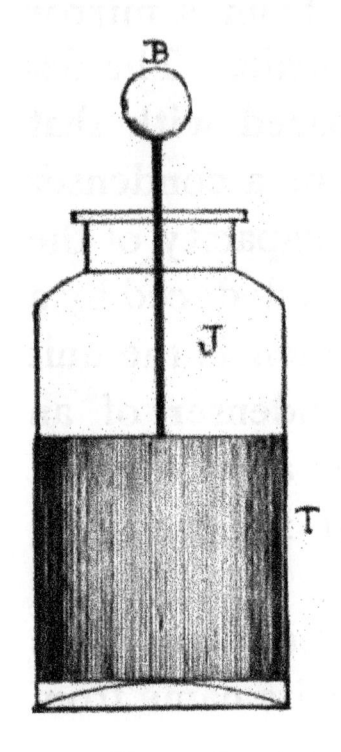

FIG. 16.

A condenser can be easily made as follows (Fig. 17):

Procure a clear glass plate, *G*, free from flaws, 11 inches square by $\frac{3}{32}$ inch thick. Give this a good coating of shellac varnish all over, sides and edges. Cut out of smooth tinfoil two sheets, *T*, 8 inches square, and round off the corners with a pair of shears. There must be no sharp corners, projections, or angles to induce leakage. Lay the glass plate on a sheet of paper, and mark its outline thereon with a pencil; then remove it and substitute a sheet of the tinfoil, and mark that. This will enable you to centre the foil. Give one side of the glass plate another coat of varnish, and so lay it on the paper that its outline coincides with the pencil outline. When the varnish has partly dried take a sheet of the trimmed foil, and by observing the pencilled marks you can lay it on the varnished plate exactly in the centre. Lay down the top edge first along this line, and carefully deposit the remainder of the foil in place. Next, with a flat brush full of varnish go over the plate, pressing out any air bubbles, and ensuring both a flat and a well-varnished surface. When this is dry,

FIG. 17.

turn over the plate and repeat the operation on the other side.

If desired, a metal hemisphere of at least an inch in diameter may be attached with varnish, first scraping the foil to make a contact. The whole plate can be swung in a cradle of two silk threads, laid on a glass tumbler, or mounted on end in a shellacked block of wood.

A strip of tinfoil, S, attached at the corner can be used as a connector. The plates must be joined in the following manner when two or more are used in conjunction, and a quantity of current is desired. They should be placed so the connecting strips project alternately from each side (Fig. 18), and all on each side joined so as to leave two terminals, one to the 1, 3, 5 plates, the other to the 2, 4, 6 plates, and so on, which, when joined, will have the same effect as would result from the use of two large plates of the same total area. The nearer the plates are together the greater capacity they will have, always supposing the insulation is good, the insulation being known as the dielectric. Another good method, when a high quality of glass can be procured, is to lay the tinfoil on the plates without varnish, piling one on top of the other, tinfoil and glass alternately, and clamping the whole securely, laying a piece of cloth top and bottom to avoid cracking the glass from the pressure. This must be kept from moisture; a strip of paraffined paper stuck along the edges and extra paraffin run on will answer very well.

FIG. 18.

In constructing these glass condensers, they must be designed to correspond with the coil with which they are to be charged. In the foregoing description we have allowed a margin of 1½ inches of glass around the foil coatings. This will make 3 inches as the maximum distance between the coatings. Although a 2-inch spark from the coil would not jump this interval, a certain discharge will take place, and the less this occurs, the more serviceable the condenser will be. Therefore a greater margin should be allowed for a longer spark than 2 inches.

In the commercial condenser for telephone and telegraph use, paraffin and paper are substituted for glass, as will be described later. Heavy paraffin oil gives excellent results, but its fluidity is disadvantageous.

There is no valid reason why paraffin could not be used on the glass plate condensers, care being observed that it is free from dirt and metallic chips. In fact, the space between the glass plates of the multiplate condenser may be filled in with paraffin, and thereby exclude the air. Only a condenser so built up is not convenient to take apart for experimental purposes.

The foregoing description of a glass insulated condenser was written with the assumption that a good quality of glass be used. But the ordinary window glass is generally useless, and paraffined paper is preferable. The quality of glass known as "hard flint glass" is best, the superior qualities being imported from Europe. This latter is used in the manufacture of the standard Leyden jar for lecture purposes.

Were it not for its cost, the finest dielectric we could use would be sheet mica. Unfortunately sheet mica over 3 inches square is expensive, and becomes rapidly more so as it becomes larger.

Standard condensers for testing are made with mica carefully selected, and retain the charge for the maximum length of time. The built-up mica condenser is immersed in molten paraffin until the same has permeated the sheets, and then the complete mass is put under a pressure until the paraffin is well set.

Paper Condenser.

The paper used in the manufacture of the commercial form is a special thin, tough linen paper carefully selected, sheet by sheet, to avoid pin-holes or flaws, and kept in an oven until used to ensure absolute dryness.

When this cannot be procured, use thin unsized writing paper of a good quality, well dried, and absolutely clean. As an example of the necessity of cleanliness, a light lead-pencil mark would serve to conduct the current entirely from a charged sheet to wherever it terminated, and if suitably located, utterly destroy the usefulness of the apparatus. Ink, which most generally contains iron, will cause trouble, and although some cheap foreign condensers are built up of old ledger pages, yet their efficiency is very uncertain.

The paper used in commercial condensers is from four to seven thousandths of an inch in thickness.

Condensers.

SERIES.

The smaller the amount of surface the less will be the capacity, but the quicker the discharge. The apparatus heretofore mentioned has had the alternate plates connected together in two series, presenting a large surface and rendering a large amount of current. A condenser so made will have a low voltage or potential, but is not so liable to leakage as one made to render a high potential. The multiple condenser of a large capacity will hardly discharge and spark over an air gap requiring a contact of the two electrodes. But a smaller one, consisting only of a single pair of small plates, will spark across quite a considerable air gap.

A number of charged condensers may be put in series, and the resultant potential thereby increased. Cut a number of pieces of paper of the desired size, say 6 inches square, and a number of sheets of foil 3 inches square. Round off the corners of the foil and build up first a sheet of paper, then a sheet of foil in its centre, then another paper and another foil sheet, and so on. There is to be no connection from sheet to sheet, only the inductive action of one on its neighbor. The foil must be considerably smaller than the paper in this construction, owing to the greater tendency to discharge round the edges of the sheets, owing to the greater potential of the current.

When the requisite number of sheets have been built up, leave a sheet of foil top and bottom for connection, tie between two pieces of stout card or board, and immerse in the molten paraffin. When thoroughly soaked, remove and put under pressure until cold. It will be found undesirable to make these with more than a dozen pairs of sheets, but to make a number of blocks of that number for ready service.

Fig. 19 shows the arrangement of the apparatus to charge a Leyden jar, the plate form being connected in a similar

Condensers. 65

manner. The jar is stood upon an insulating support—a dry tumbler will answer—

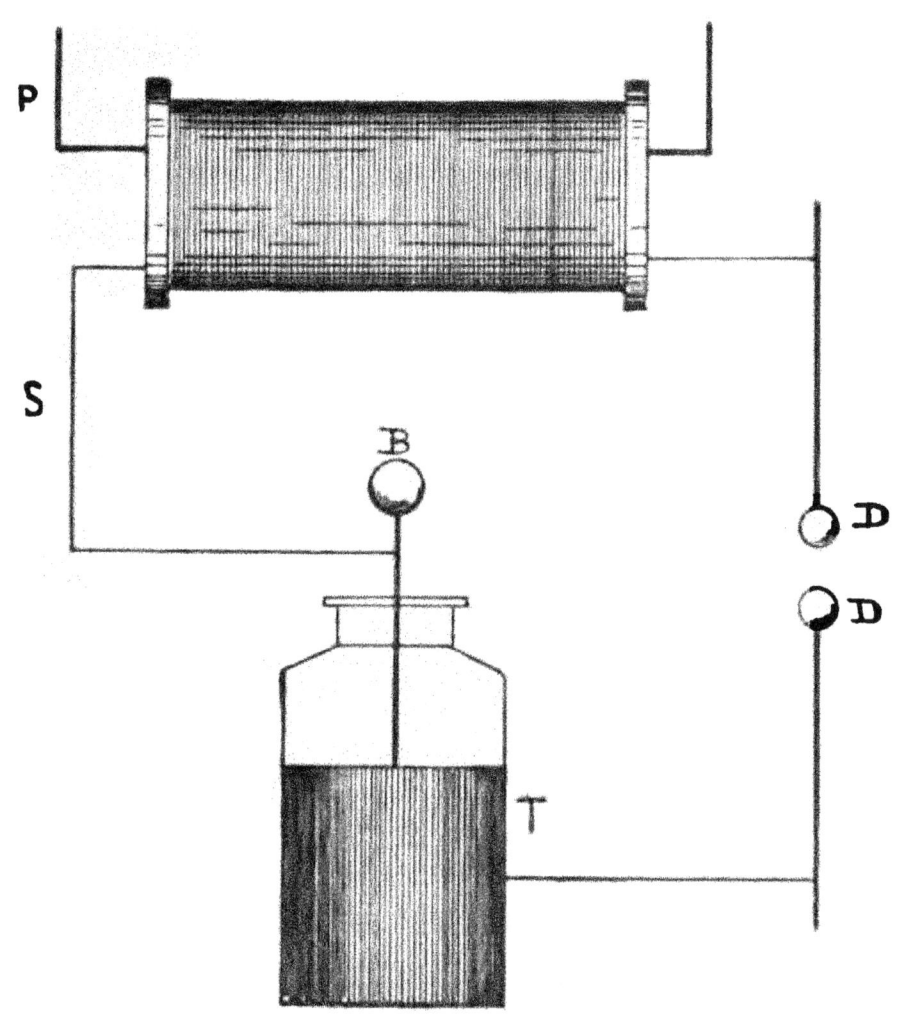

FIG. 19.

with the ball *B* connected to one pole of the coil. From the outside tinfoil coating

T a wire runs to the discharger *D D*, which is in circuit with the secondary coil, *S*. The discharger balls *D D* are carefully approximated until the spark just passes, this latter point being of great importance. Were the discharger balls too near, the spark would probably pierce the dielectric of the condenser, therefore the balls should be carefully *brought near* to each other until the exact distance is found. Even if the insulation of the condenser were not pierced, yet a path would probably be opened through which some succeeding discharge would pass, and ruin the instrument.

Another method of charging is to leave an air gap at *B;* then there is not much liability of the condenser discharging back through the coil—an undesirable event, as it would most likely perforate the insulation of the coil.

In designing or using any apparatus intended to hold a charge of high potential, it must be kept in mind how readily points or

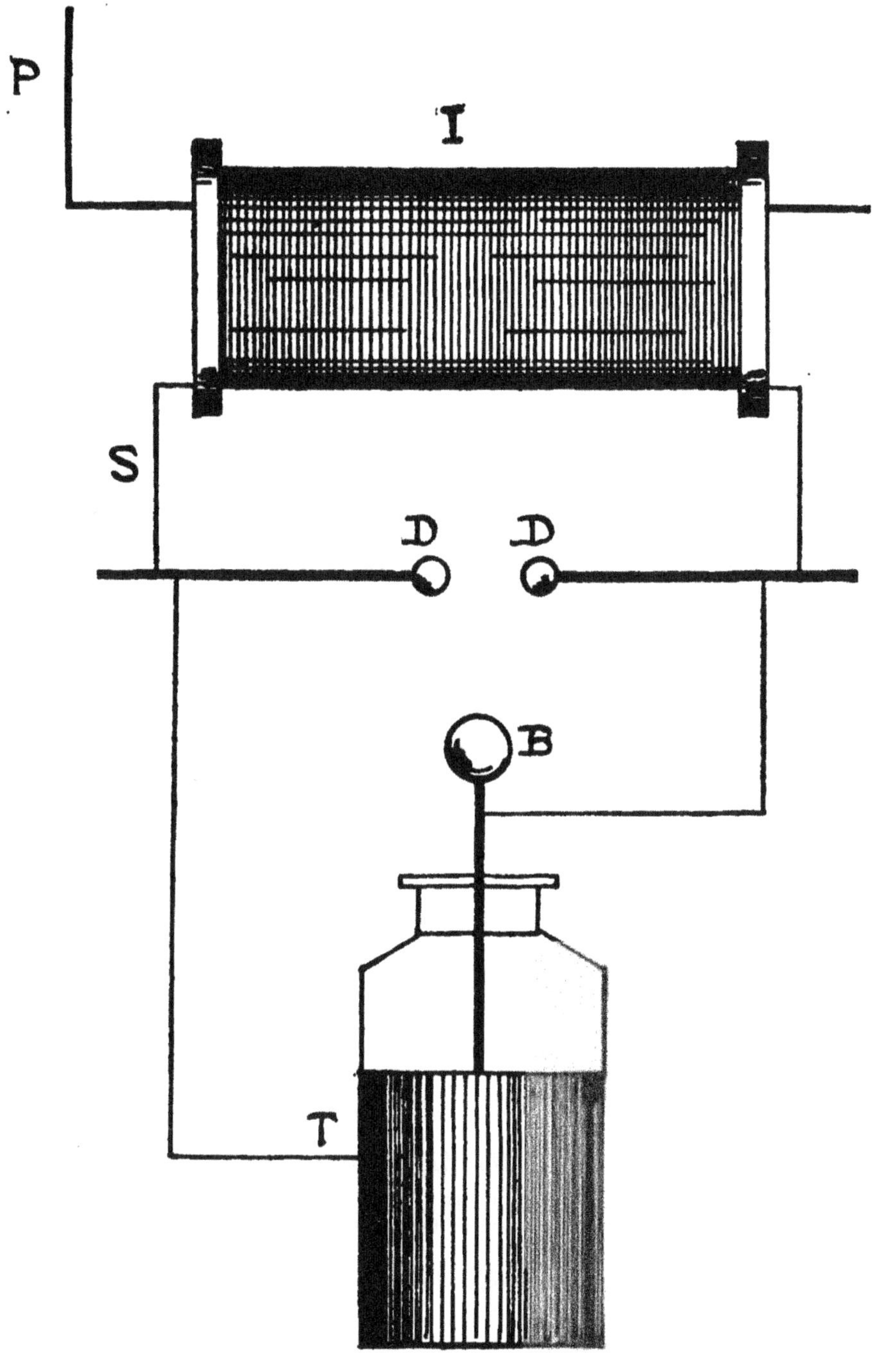

FIG. 20.

sharp edges serve to allow the current to pass off—we might almost say evaporate. Given two bodies, one a globe and the other a rectangular block, each well insulated from the earth or any other large body, and the globe would be found to hold its charge long after the block had dissipated all trace of the charge given to it. Therefore round off every edge and angle, projection or point.

In making handles, supports, or any work requiring an intervening high insulation, hard rubber is preferable to glass where there is liability to moisture. When the apparatus is as shown in Fig. 20, the condenser is alternately charged and discharged with a loud noise, the vivid sparks passing across the discharger balls $D\ D$ possessing great deflagratory powers.

In experimenting with a Ruhmkorff coil it is not advisable to leave the instrument working while the secondary terminals are beyond sparking distance, as there is a great strain on the secondary insulation.

Nor is it wise to use only one electrode in an experiment, unless the other is connected to some apparatus of an approximate capacity to that at the other, for the foregoing reason.

CHAPTER V.

EXPERIMENTS.

THE luminous effects that can be obtained by means of a Ruhmkorff coil are exceedingly beautiful and instructive. The simplest experiment of this nature is the production of the spark consequent on the approximation of the electrodes attached to the secondary coil. This spark can be varied in both length, intensity, or shape by the form and nature of the substances between which it is permitted to pass. Attach to each end of the discharger a fine steel needle, and bring them together until the spark jumps from one to the other. A long thin snapping spark will pass, which, however, appears to be trying to take any but a straight path across the air gap. The peculiar crookedness of this, as in a lightning flash, is credited to the fact of particles of matter floating in the air conducting the current better than the pure air. The curious odor noticed in these discharges, as, in fact, in the working of all high-tension apparatus, is ozone—O_3, triatomic oxygen. This gas, so noticeable after a thunderstorm, has a powerful effect on the mucous membranes of the throat and nasal passages, and must be inhaled with caution. It is being used by the medical profession for the destruction of germs and for general therapeutic service.

Substitute pieces of fine iron wire for the needles, and bring the ends together about one quarter the distance through which the normal spark will pass. The spark will be found to have changed its appearance, now being thick and redder, or, rather, of a deep yellow, and to possess vast heating qualities.

The iron wire will melt at one electrode, and if the other be examined it will be perceived that it has not even become warm. The cold wire will be the one connected to the positive pole of the coil.

Connecting the poles together with a piece of very fine iron wire will result in the deflagration of the wire in a vivid light.

The short thick spark is termed the calorific spark, and believed to possess its yellow color from the combustion of the sodium in the air. This spark will easily ignite a piece of paper held in its path.

Take a sheet of hard rubber and breathe on its surface; lay a wire from each pole of the secondary to points on the sheet, about twice as far apart as the spark would pass over in the air. The electric current will strive to complete its circuit; streams of violet light forming a perfect network will issue from each pole, until, provided the rubber is sufficiently damp, they will unite in a spark far exceeding its normal length in the air. It is curious to watch how the streams branch out from these two points, and how persistently they strive to meet each other. Scatter some finely powdered carbon on this sheet (crushed lead-pencil or electric light carbon is good material). The points may now be removed to still further distant places, and yet the current will work across. Each particle of carbon seems to be provided with innumerable scintillating diamonds, so sparkling is this effect.

Hard rubber is not absolutely necessary for these experiments; glass will do, but the black background of the rubber intensifies the luminosity of the discharges. Take a teaspoonful of powdered carbon and scatter it between the points on the rubber, so that the spark can find a ready path, evidenced by but little visible light. It will be seen that this powder is blown away from one electrode after a few minutes, leaving the latter in the centre of a clear space, but at the other electrode not much disturbed.

Bring the points so close to one another

that the spark becomes short and fat; soon the carbon will commence to burn, forming a veritable arc light. Take two pointed lead-pencils and wrap a few turns of wire from the electrodes round the blunt ends of them; bring the pointed ends together, and an arc will soon be established; but at various points where the wire is wrapped the current will burn through the wood, and a number of incandescent points will ensue.

In these experiments on the rubber sheet it will be noticed that the spark acts as it does in the air, inasmuch as it does not take a direct path, but jumps in an irregular track from point to point.

If two small metal balls be substituted (Fig. 21) for the points between which the sparks be passing, it will be noted that the sparks do not pass through so great an air gap as before, or even as rapidly.

The spark between two balls is much noisier than that passing between points, and if the balls be of about 1 inch in

Experiments. 75

diameter, a curious effect ensues on the passage of the current (Fig. 22). This

FIG. 21.   FIG. 22.

effect has been likened to a stream of water issuing from a horizontal nozzle into a cavity when the nozzle is moved up and down slowly in the space of a few inches.

THE LUMINOUS PANE.

This easily made exhibit (Fig. 23) is one that is susceptible of quite a number of applications. In its simple form it is but an enlarged version of the rubber sheet scattered with carbon dust. The old way to make it was to take a plate of glass and ce-

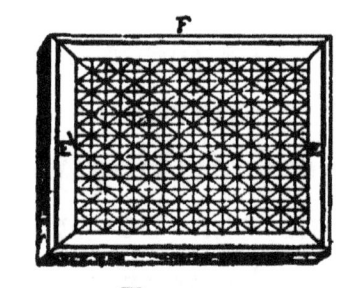

FIG. 23

ment on one face of it a sheet of tinfoil, using shellac varnish preferably. When dry, the tinfoil was scored across and across in such manner as to divide it up into little squares or diamonds. When the current was applied to each end of the plate, the spark divided into innumerable little ones; between each bit of tinfoil and its neighbors there would be many little sparks, and the effect was very pretty, somewhat as was described before when the carbon dust was strewn between the electrodes. It is more easily and quickly prepared by giving a sheet of glass a coating of shellac varnish, and then sparingly dusting any powdered conductor over its surface, using perhaps carbon dust or filings of metal. By cutting out a stencil from a piece of thin card and laying it over the sparkling plate, the design shows out very strikingly, and various designs in stencils can be prepared, different powdered conductors giving different colored sparks.

A long glass tube moistened inside with mucilage or shellac varnish and then having some conducting dust shaken through will also give quite a pleasing effect.

Luminous Designs.

Coat one side of a glass plate with tinfoil, leaving an attached strip for connection. Shellac a piece of paper of a size corresponding to the design to be rendered luminous. When the shellac has dried so far as to become "tacky," lay a sheet of foil on it and press it down evenly all over.

Then draw on the paper a design that can be readily cut out. Use a pair of scissors or a very sharp knife. If the latter, lay the sheet on a piece of glass; but there is a greater tendency to tear the design when a knife is used if an unpractised hand wields it.

This design may either be stuck on to the plain side of the glass plate with varnish or simply laid on (Fig. 24). Connect one secondary wire to the foil coating of the plate and the other to the design. This must be shown in the dark, and the luminosity will not be strikingly apparent until the eyes become accustomed to the darkness—that is, when the room has been previously lighted.

One of the most beautiful and easily obtained phenomena of the high-tension discharge is the "electric brush" (Fig. 25).

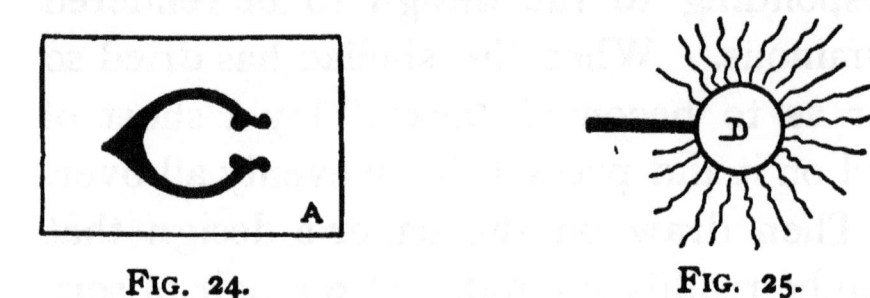

FIG. 24.   FIG. 25.

This occurs when the secondary electrodes of the coil are too far apart to allow of the free passage of the spark, and can only be seen at its best in a perfectly dark place. The ball tips before mentioned show this brush very plainly, or two sheets of tinfoil in circuit hung far enough apart to prevent vivid sparking will cause this so-called "silent" discharge. This latter arrangement should not be used for over fifteen minutes, as the ozone which is liberated in large quantities will affect those persons in the vicinity.

In fact, when a rapid vibrator is being used with the coil, the leading wires from the secondary terminals present this brush appearance, the curious threads of light resembling luminous hairs waving in the air. The more rapid the vibrations the more prominent the brush effect, as will be seen in the Tesla coils. The positive ball of the discharger shows the brush as a spreading mass of luminous threads reaching out toward the negative ball, which latter resembles a star, as in the figure.

The intensely disruptive power of the long spark is readily shown by its power to perforate substances, but great care must be taken that the secondary wires of a coil are led away from the body of the coil. A good plan is to hang two silk cords or stout threads from the ceiling, to which the secondary wires may be attached and kept in sight when experimenting at any distance from the coil.

To pierce a piece of thin glass, take two

lumps of paraffin about the size of a walnut, and, warming them and the glass sheet, stick them on opposite sides of the glass facing each other. Then warm the ends of the two pointed wires and thrust them into the lumps of paraffin, that they terminate on the glass surface directly opposite each other. On connecting these to the secondary coil a few impulses to the contact breaker will start an electric discharge sufficient to pierce the glass if the thickness be proportioned to the power of the apparatus. The great Spottiswood coil pierced a block of glass 6 inches in thickness.

There is, however, a certain element of danger to the secondary insulation in performing this experiment.

CHAPTER VI.

SPECTRUM ANALYSIS.

IF a metal or the salt of a metal be burned in a flame it imparts to the flame a distinctive color; table salt thrown into the fire burns with a yellowish flame, denoting the presence of sodium, and a greenish tint, indicating the combustion of chlorine. Violet flames accompany the burning of the salts of potassium, and barium burns green. Lithium and strontium give a red hue. But to be ordinarily perceptible, the salts require for the most part to be present in considerable quantities. By the use of the spectroscope, however, extremely small proportions of these metals and salts can be readily detected and classified.

If a beam of light be transmitted

through a prism of glass the rays are decomposed, and what is known as a spectrum is formed (Fig. 26). The most generally observed spectrum is the rainbow. When the light from a flame in which is burning some suitable substance be trans-

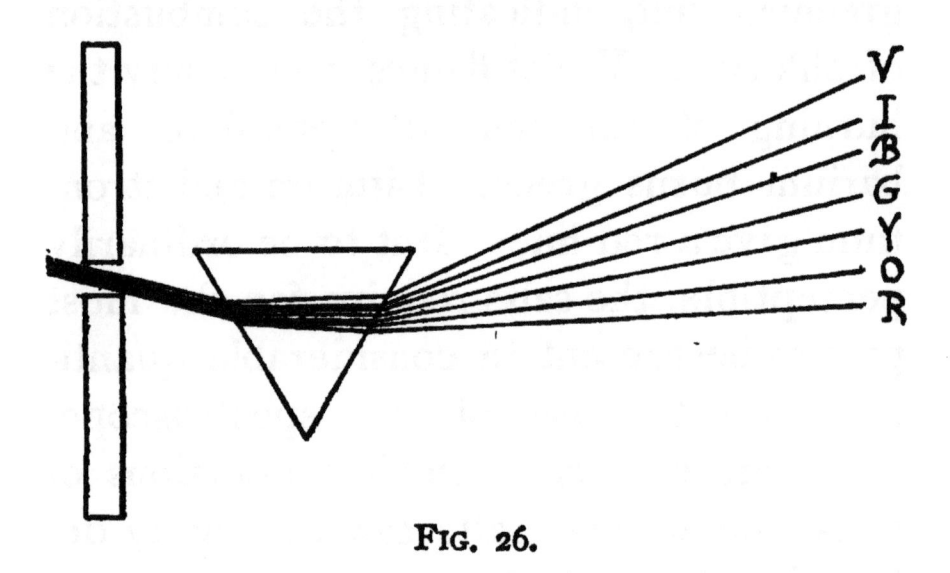

FIG. 26.

mitted through the prism, the color which predominates in the flame will predominate in its spectrum. The combination of a prism and tubes for observing these effects is a spectroscope (Fig. 27). The short fat spark from the Rhumkorff coil is most useful in this work. The electrodes are provided with a portion of the substance to be examined, and the spark is

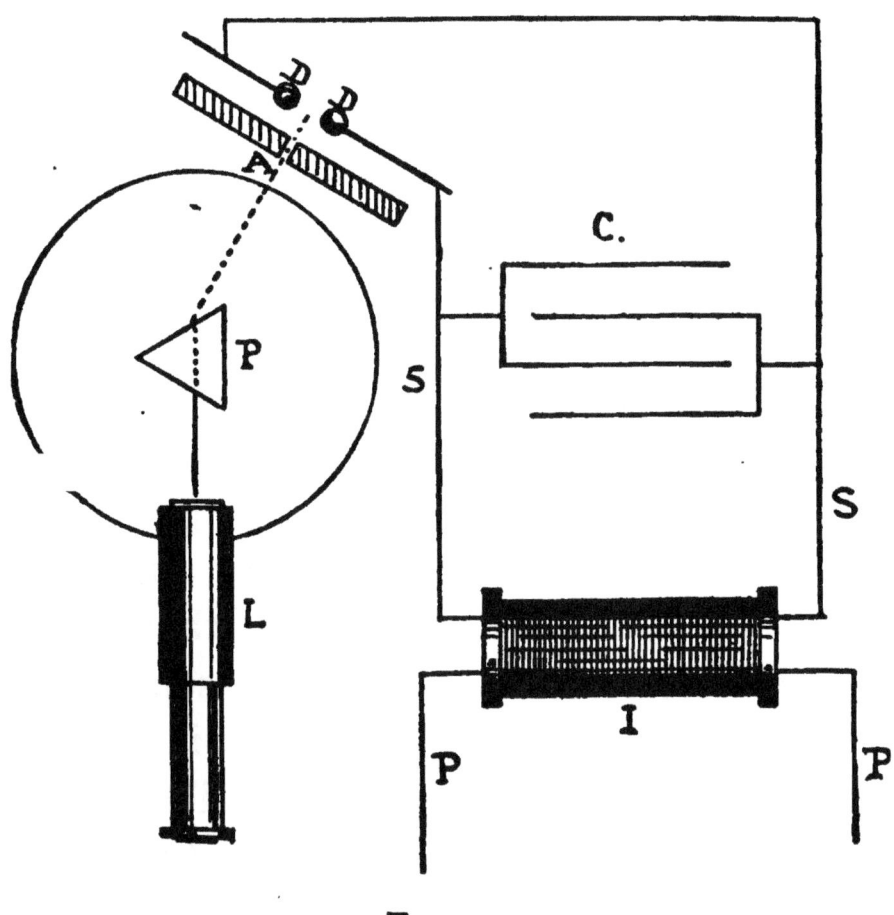

FIG. 27.

passed and viewed through the **spectroscope**.

The spectroscope is shown in connection with the coil in Fig. 25. *A* is the aperture in the screen through which the rays from the metal burning at the discharger balls DD passes. The lens at L is used to view these rays after they have been decomposed by the prism P, which, as well as the lens, can be rotated. I is the coil, PP the primary and SS the secondary wires, C being a condenser bridged across the circuit.

The screen should be pierced by a very narrow aperture, A, and be placed at a considerable distance from the prism P, that the rays issuing through the aperture may not strike the prism until they have widely diverged and become separated from each other. The aperture is practically formed of perfectly parallel knife edges, forming a slit not exceeding one hundredth of an inch in width.

The colored spaces in the solar spectrum do not occupy an equal extent of area; the violet is the most extended, the orange the least. The proportion is in three hundred parts: Violet, 80; green,

60; yellow, 48; red, 45; indigo, 40; orange, 27.

The solar rays exhibit on careful examination dark lines crossing the spectrum at right angles to the order of the colors, and always occupying the same relative positions. These are called Fraunhofer's lines.

If, however, the spectra of metals, gases, and other elements be examined they will be found to present certain characteristic *bright* lines, the body of the spectrum being often feeble or entirely dark. The spectrum of hydrogen gives two very bright lines of red and orange.

An extremely minute quantity of an element is necessary to give distinct lines. Sodium gives a single or double line of yellow light in a position agreeing with that of the orange rays in the solar spectrum.

Potassium gives a red line in the red end and a violet line in the violet end of the solar spectrum. Strontium presents eight bright lines; calcium gives mainly one broad green band and one bright orange band.

In practical work with the spectroscope a solar spectrum is often arranged that it can be used as a comparison with the spectrum being investigated, one spectrum being formed above the other, and the observation made as to which lines coincide. Iron gives nearly sixty bright lines coinciding with the same number of dark lines of the solar spectrum.

The violet rays of the solar spectrum are the rays which possess the maximum chemical action, the yellow the maximum light effect, the red the maximum heating effect. Beyond the violet band of the spectrum exist certain rays termed the invisible rays or ultra-violet rays, which in themselves are not luminous. Their vibratory rate is higher and their wave length shorter than the violet rays, according to the most generally accepted theory of light. These rays, when passed through certain substances, suffer a change and become visible in a luminous state of the substance, which luminosity is termed fluorescence.

The bright yellow line of sodium in the orange rays is found in nearly all spectra, owing to its extensive diffusion in the atmosphere.

Tesla has succeeded in producing electric waves of length approximating to those of white light, which appear to have very little heat. The ideal light is that which shows no heat and does not liberate noxious gases in the air, and were it not for its feeble luminosity, the light of the electric spark passing through a carbonic acid vacuum would approximate this most nearly.

The present mode of obtaining light—that of raising to a high temperature some substance or collection of particles—seems certainly somewhat antiquated. The following notes may be of interest and assistance in researches bearing on the lighting question.

Solid bodies, when heated, show a red glow in daylight at an elevation of temperature corresponding to 1000° Fahr.

Temperature, degrees F.	Color of Substance.
1000	Red.
1200	Orange.
1300	Yellow.
1500	Blue.
1700	Indigo.
2000	Violet.
2130	All colors—*i.e.*, white.

The number of vibrations per second necessary for the production of light, and the velocity of light being determined, the calculation of the wave lengths of the colored rays becomes possible.

The following table (Sprague) shows this in ten-millionths of a millimetre (a millimetre = .039 inch) measured in the dark lines of the solar spectrum, from red to violet:

Orange =	6.88
Orange, Higher =	6.56

Yellow = ... 5.89
Green = ... 5.26
Blue = ... 4.84
Blue, Higher = ... 4.29
Violet = ... 3.93

CHAPTER VII.

CURRENTS IN VACUO.

Notwithstanding it requires an intensely high potential to enable the current to jump an air gap of 1 inch, the same potential will produce a luminous discharge through exhausted glass tubes aggregating 8 feet or even more.

But the exhaustion can be carried so far that there is no apparent discharge; and, on the contrary, air at as high a pressure as 600 pounds per square inch will resist the passage of the spark over an extremely short space. If the tubes be filled with various gases and then partially exhausted, the length of tube through which the luminous discharge will pass varies with the gas, becoming shorter in the following order: Hydrogen, nitrogen, air, oxygen, and carbonic acid—the shortest.

Before detailing some of the more striking phenomena connected with high-tension discharges in vacuo, a description of a few forms of simple mercurial air pumps will be serviceable.

Fig. 28: If a glass tube, *F*, stopped at one end, 3 feet long or over, be filled with mercury and the open end immersed in a vessel of mercury, *T*, the column of metal in the tube will sink until it attains a height, *M*, of about 30 inches, varying according to the condition of the atmosphere.

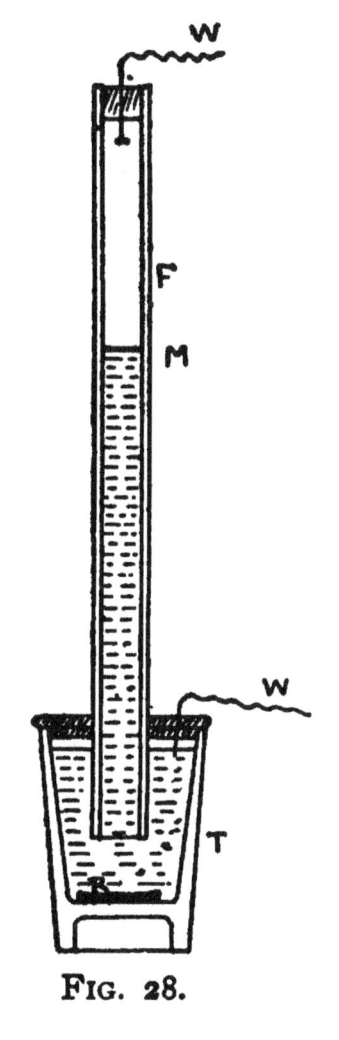

FIG. 28.

The space between the mercury column and the top of the tube will be a fairly good vacuum. This fact was noted many years ago, and the gradual evolution of the mercurial air-pump based on this result can be followed in the articles on the mercurial air-pump by Silvanus P. Thompson, read before the Society of Arts, England, some years ago.

Geissler, the first manufacturer of the "Geissler" or vacuum tube for electrical research, seeing the inconvenience of the above-described operation and the meagre results obtained, invented the pump called by his name (Fig. 29).

FE is a stout glass tube some 3 feet long, having a bulb, B, at its upper extremity, and a rubber tube, S, attached to the curved end. A reservoir of mercury, R, connects with this rubber tube, and a special glass tap is fixed in the upper end of the glass tube at E, beyond which tap being the point of attachment for the object to be exhausted. The operation is as follows: On turning the tap a part of the way it allows a passage between the tube FE and the atmosphere. The reservoir

R is then raised until the mercury flows into the bulb and up the tube to the tap. The tap is then turned a fraction, and the communication with the air is shut off and opened between the object to be exhausted and the tube FE. The reservoir is then lowered and the mercury falls, drawing down the air from the object into the tube. The tap is then turned as in the first place, and the reservoir R raised, when the air drawn into the tube is forced out by the rising column of metal. This operation being repeated many times, withdraws nearly all the air from the object—in fact, makes a fairly good vacuum. This pump has been much modified from the simple form described.

The form of pump most used in the United States lamp factories is based on the application of the piston-like action of a quantity of mercury dropping down a tube. This is known as the Sprengel pump, after the inventor.

Fig. 30: F is a stout glass tube about

FIG. 29.

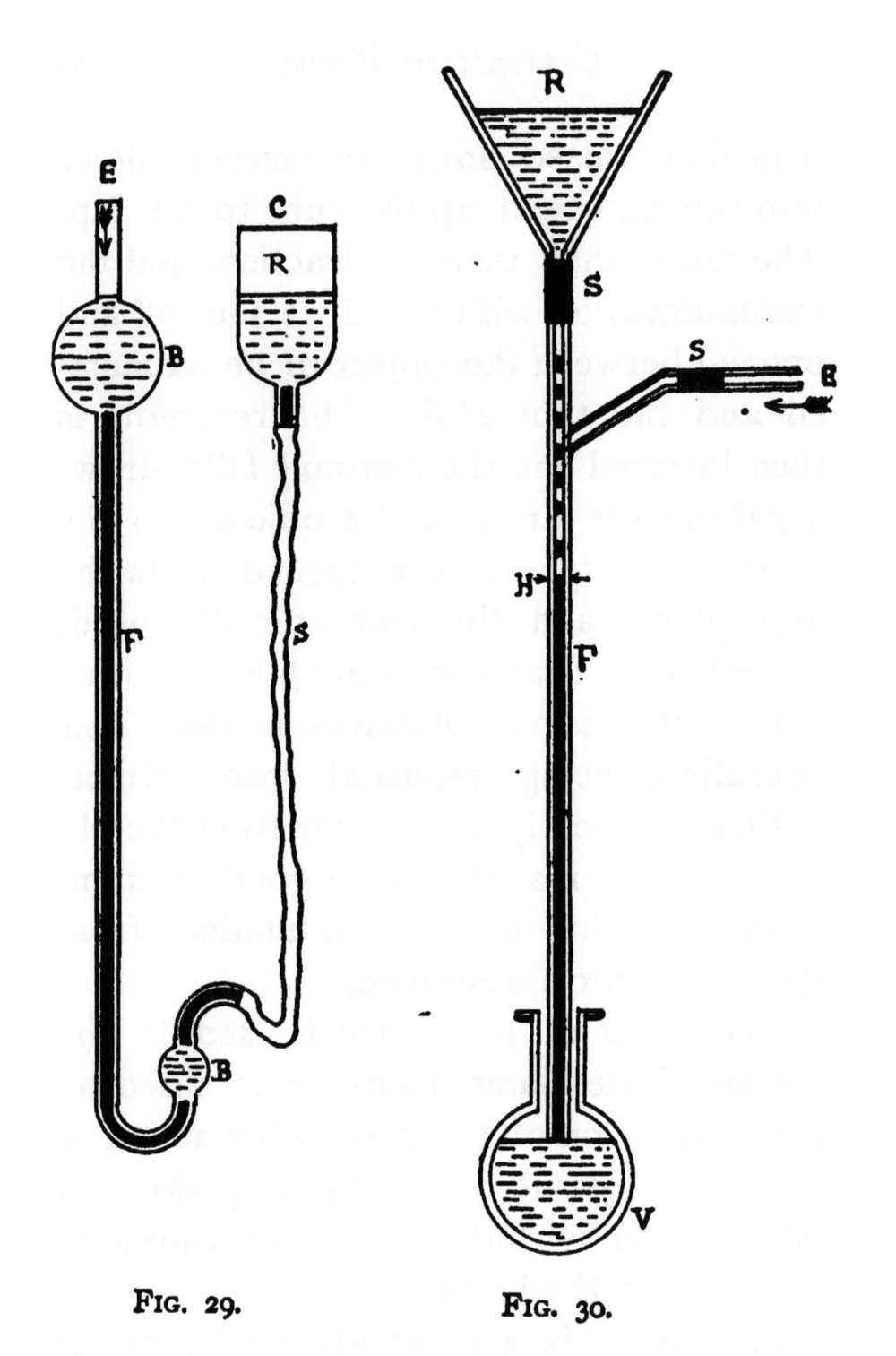

FIG. 30.

40 inches long by one-twelfth of an inch internal diameter, carrying the reservoir funnel R at the top, a piece of soft rubber tubing, S, nipped by a pinch-cock being interposed to admit of the regulation of the mercurial drops. The lower end of this "fall tube," as it is called, is immersed in mercury contained in a suitable vessel, V, a branch tube being blown or cemented into the fall tube to admit of the connection of the object to be exhausted at E. S is another piece of rubber tubing with a pinch-cock regulation. The point H is the normal barometric height of the mercury—about 30 inches. On attaching a bulb, for example, at E, and regulating the pinch-cock at the top of the fall tube F, a succession of drops of mercury falls down the tube, each drop acting as a piston to drive the air before it, sucking the same from the bulb, and forcing it down through the tube and vessel out into the atmosphere.

On its first being set into operation, the

cushions of air between the drops silence their fall; but as a higher degree of rarefaction occurs, the air cushions become insufficient, and the drops fall with a sharp click on the top of the barometric column.

One great disadvantage in this form of pump is the tendency to fracture of the glass tube that is manifested by the concussion of the drops of mercury at the barometric height. However, this has to a certain extent been obviated in later forms of this useful and efficient pump.

For many electrical experiments, the simple exhaust tube (Fig. 28) mentioned at the beginning of the article will be found very satisfactory. The top end need not necessarily be sealed off with glass, a cork having a wire, W, run through for connection being driven in, and a coat of paraffin or one of the cements mentioned in a later chapter be laid on.

The second electrical connection is made by a wire dipping in the tumbler of mercury.

DISCHARGES IN VACUO.

In a simple glass tube having two wires carrying balls inserted through its ends, from which the air has been partially exhausted, the study of the changes shown by the passage of the spark is extremely interesting. Before the commencement of exhaustion no luminous effect can be discerned; at a low degree of exhaustion a luminosity appears between the ends of the wires, the negative pole being surrounded by a violet glow and a larger pear-shaped red discharge from the positive. An interval near the negative electrode is in darkness, widening as the exhaustion progresses. When the degree of exhaustion is very high, a series of arches concentric with the positive ball appear and become broader and more distinct as the rarefaction progresses. The arches or bands are called striæ, and are most distinct when the tube is made in the form of a narrow cylinder, with a bulb at each end.

Carbonic acid gas vacua give the best results. If the finger be placed on the bulb at either end a luminous spot appears, and by using a very rapid contact breaker in the primary circuit, the luminous discharges become highly sensitive, being diverted from their regular path on the approach of the hand, a magnet, or a grounded wire. An extended treatment of these phenomena would be out of place here, but can be found in nearly all comprehensive works on electricity.

If an incandescent-lamp bulb be held in the hand and one end be brought near to a terminal of the coil, a beautiful bluish light appears. The carbon filament, if long, and not held by its loop, becomes electrified and oscillates, often giving out a clear, high, bell-like sound as it strikes the glass. Particles of carbon deposited on the glass during the burning of the lamp, shown in daylight as a blackening deposit, generally show little sparks, like stars scattered over the inside of the globe.

A vacuum tube will phosphoresce if held in the hand near a secondary terminal, or even if laid on the table near the coil, and will light quite brilliantly if one end be held against a terminal. This latter method, however, is generally inconvenient, as a certain amount of physical pain ensues from the discharge into the skin.

Different gases in the tubes give characteristic colors. In carbonic acid gas the whitish green hue prevails; in hydrogen, white and red; in nitrogen, orange yellow. The characteristic spectra are given by the gases in the tubes, and can be readily examined in the spectroscope. But there is sometimes a slight variation in these colors, dependent upon changes in the current.

In many Geissler tubes, a portion of the bulbs is made of uranium glass. On the passage of the spark in the tube this glass glows with a magnificent emerald green hue. Other tubes are constructed with

100   *Currents in Vacuo.*

an outside enveloping glass tube fitted with a corked orifice into which can be poured different solutions.

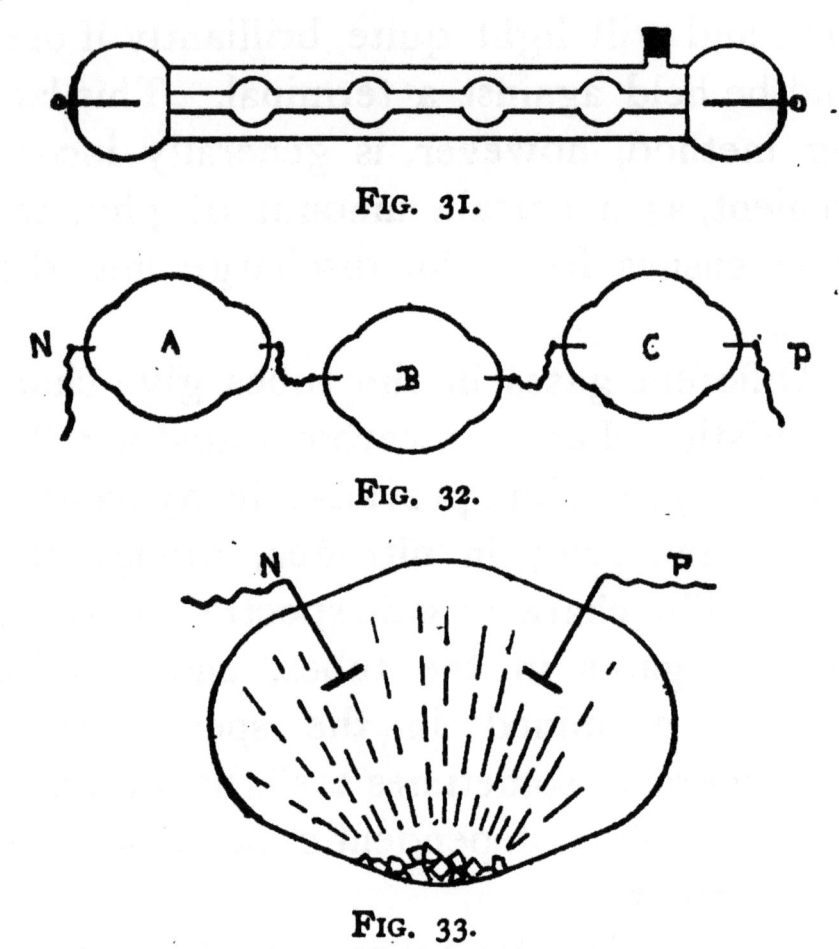

FIG. 31.

FIG. 32.

FIG. 33.

Fig. 31 shows a solution tube to be filled with solution of sulphate of quinine, etc.

Fig. 32 shows three exhausted tubes arranged in series.

A is of uranium glass, and glows dark green ; *B* of English glass, showing a blue hue, and *C* of soft German glass, glowing with a bright apple-green tint.

Crystals of nitrate of calcium, nitrate of silver, benzoic acid, tungstate of calcium, lithia benzoate, sodium salicylate, zinc sulphide, and acetate of zinc fluoresce.

Fig. 33 is a highly exhausted tube, having at its lowest part a few pieces of ruby. When the secondary current is turned on at *P* and *N* the rubies shine with a brilliant rich red, as if they were glowing hot.

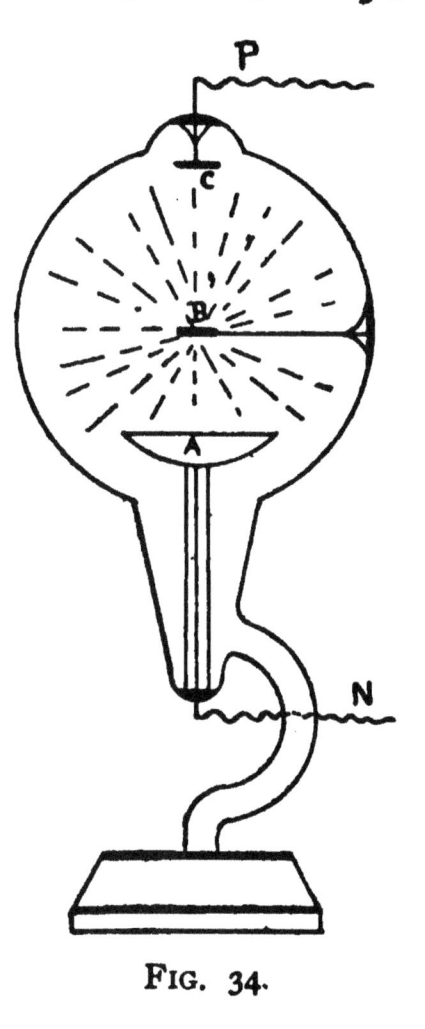

Fig. 34.

Fig. 34 shows the tube to exhibit the

effect resulting from focussing the electric rays on a piece of iridio-platinum at B.

The cup A forms the negative pole; the metal disk C, the positive.

On increasing the intensity of the spark, the metal at B glows with extreme brilliancy, and melts if the intensity be carried too far.

CHAPTER VIII.

ROTATING EFFECTS.

ALTHOUGH the luminous discharges in the exhausted tubes are extremely beautiful, yet the effect is indescribably enhanced when the tubes are rotated. Gassiot's star was the name given to the earliest exhibit of a rotating tube carrying a luminous discharge, owing to the curious phenomenon ensuing from the interruptions of the spark. As the human retina is only capable of retaining an impression for a fraction of a second, and as the tube is only momentarily luminous during the passage of the spark, the effect of the revolving tube is that of a series of such arranged as the radii of a circle, the number apparent, being governed by the rapidity of rotation and the rate of interruption of the current.

Fig. 35 represents a form of rotating wheel which is easily made, and yet susceptible of many novel and attractive

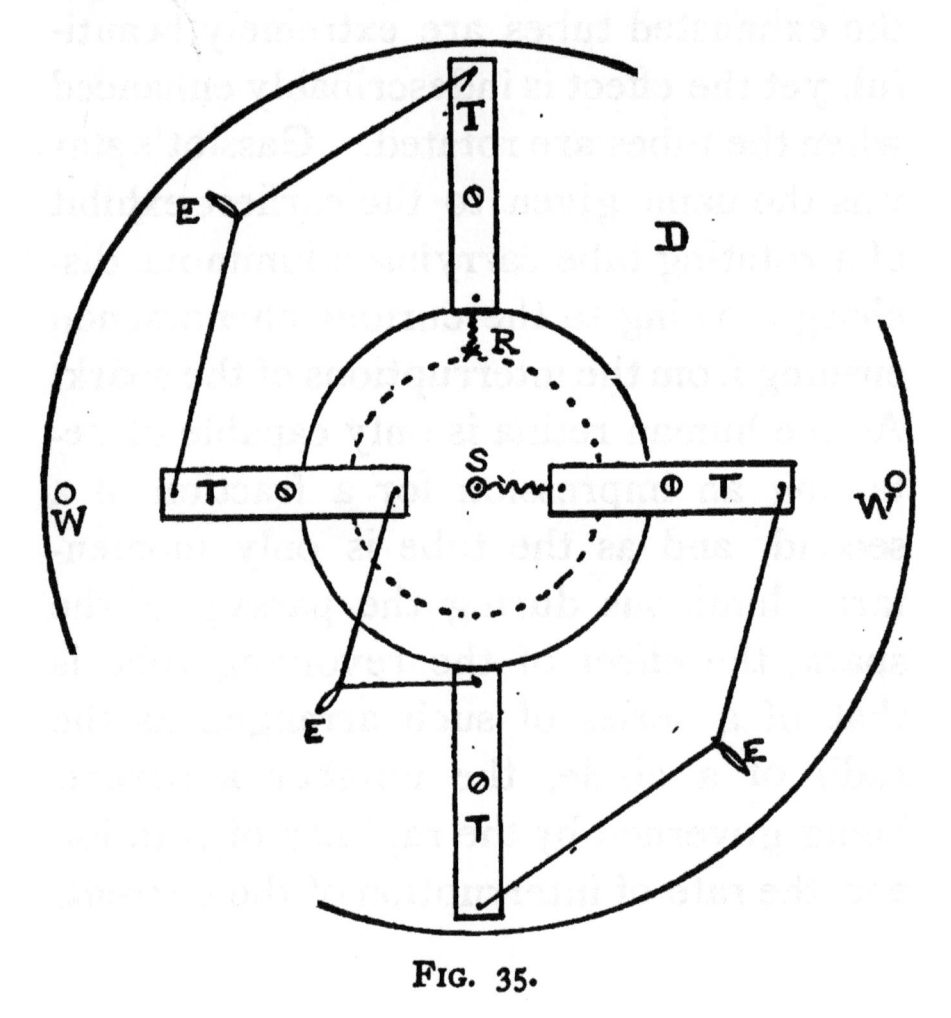

FIG. 35.

effects. Such a wheel, placed in a store window, would undoubtedly attract many persons by the beautiful variations of colored figures which it presents while in motion. And once a crowd is collected and its attention attracted to one spot, the capabilities of advertising the goods on sale are apparent.

A pasteboard or light wooden disk D, 3 feet in diameter or over, is mounted on a shaft, S, operated by an electric motor or such power as may be attainable. Upon its surface are mounted the tube-holders $T\ T\ T.T$, connected, as shown, by wires leading from the secondary of the Ruhmkorff coil. Starting at the shaft S, the circuit runs to the first tube-holder, where the continuity of the wire is broken to allow of the attachment of the vacuum tube. From the first tube-holder the wire runs in turn to each of the other three tube-holders, terminating at R, where it passes through a hole to a metal ring on the back of the disk shown by the dotted circle. This ring and the shaft are in connection with the secondary coil, by reason of its electrodes being attached to two brushes or strips of metal pressing, one on the ring, the other on the shaft; or the bearing in which the shaft turns may displace one of the brushes. *W W* are two counterbalance weights, that the wheel may run smoothly and be not affected by the irregular distribution of the tubes or its surface. *E E* are elastic bands, looped over the wire and through rings in the disk, that the wires may not be liable to touch or short circuit.

At Fig. 36 is an enlarged view of a tube-holder, although, as it is meant only as a diagram, considerable variation of design is permissible. The springs at *H H*, to which the wires run, being bent back, the metal pins *P P* may be thrust through the rings on the ends of the tube, and the elasticity and pressure of the spring will hold it in place and make the necessary contact. A wooden block, *B*, secured to the face of the disk, is provided with a thumb-screw, *S*, securing the tube-holder to it, by means of which the tube-holders may

Rotating Effects. 107

be turned a trifle upon their axes and so vary the effect of the wheel.

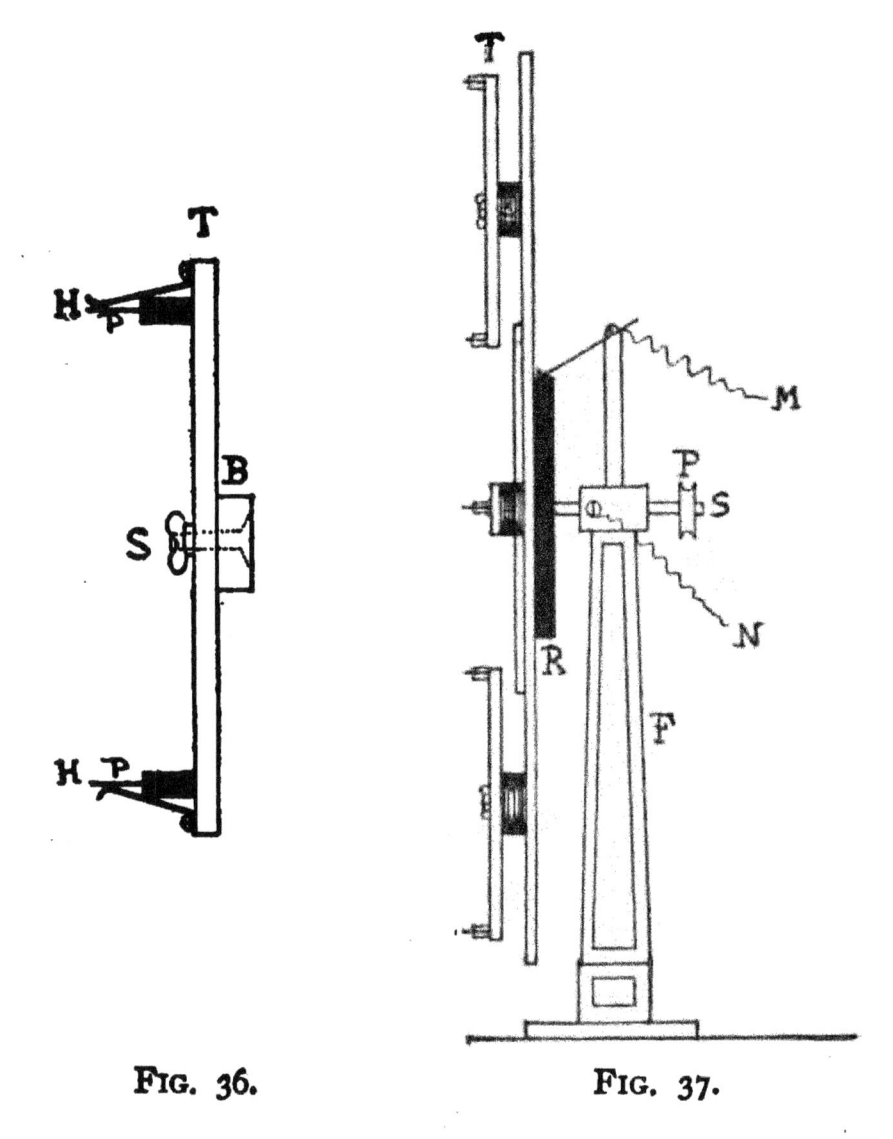

FIG. 36.

FIG. 37.

Fig. 37 is a side view of the wheel, showing one manner of mounting the disk and its connections. The same figures apply to the parts as in the preceding figure. *M N* are the wires leading to the coil, *P* is a pulley on the shaft whereby the rotary power may be applied. The wires on the face of the disk are not shown, as they would impair the clearness of the diagram unnecessarily.

The greatest danger in the operation of such a piece of apparatus will be the tendency of the high tension spark to wander where it is not wanted, and to take short but forbidden paths back to the coil. However, care and perhaps experiment will prove the remedy. It will be noticed by reference to Fig. 32 that a circle has been drawn almost bisecting two of the tubeholders. This circle represents a circle of danger, and where a thin material has been used for the disk, the disk may very well be reinforced by a piece of stouter card cemented on its face.

The disk, whether of wood or of pasteboard, must have a liberal coating of insulation, either shellac varnish, paraffin, or beeswax, and be absolutely free from unnecessary holes. Moreover, the ring R must be of such a distance from the support F, if the latter be metal, as will preclude any jumping of the spark. A Ruhmkorff coil giving upward of three quarters of an inch of spark will be large enough to operate a wheel carrying four 8-inch tubes.

The wheel may be set back in a window and surrounded by dark fabrics, or built in, as it were, in a cave of such. The judicious use of pieces of looking-glass scattered on the sides of the cave, in such manner as to reflect the light of the tubes, will enhance the effect. There is no danger of fire where ordinary care is used, as the *long* spark necessary to the production of the luminosity will hardly ignite anything but gas, unless specially arranged to do so.

Fig. 38 is a triangle formed of three Geissler tubes, and intended for rotation as a whole. *M M* are two pieces of mica or glass, to prevent any possibility of the spark jumping and short circuiting, in which event the tubes would fail to light.

This triangle is shown diagrammatically at *A B C*, Fig. 39, mounted on an insulated

FIG. 38.

rotating disk. Before commencement of rotation, and upon the current being turned on to the tubes, a simple triangle will result, but at a certain stage of rotation the Maltese cross shown is formed. A still higher rate of rotation will produce the

Rotating Effects.

double star, Fig. 40, and as the rotation and rate of vibration of the coil contact-breaker is varied, an apparently endless succession of stars or triangles appears to grow out into view.

Although Figs. 39 and 40 serve to illustrate a triangle of tubes and its varia-

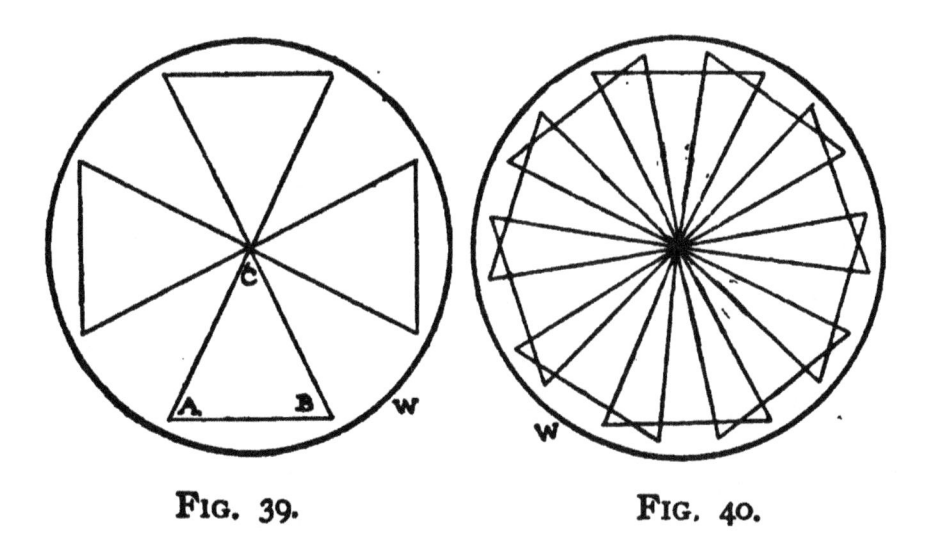

FIG. 39.   FIG. 40.

tions, a very pretty and simple effect can be obtained with it as follows: Three strips of looking-glass are cut and scratched across their silvered surface, as described for the luminous pane, Fig. 43. The current then being allowed to pass, and the wheel being rotated, the triangle acts as in the preceding paragraphs, multiplying and forming figures, which are extremely interesting to watch.

While treating on the subject of store-window attractions, a few suggestions on a display of stationary Geissler tubes may be made. Starting with the assumption that the platform on which the goods would be displayed is of wood, a very small amount of preparation is necessary. The platform is covered with a dark material free from gloss, such as canton flannel, on which the tubes are laid in any fancy pattern, or may be scattered haphazard. Fine bare wire (No. 36 B. & S. is not any too small) is run from tube to tube, using care that it does not touch itself in such manner as to short circuit the current. There is not much necessity to cover the wires, unless the rate of vibration of the contact be so rapid as to show the brush discharge from the wire strands. In a jewelry store the cylindrical portions of the tubes may be covered with strips of dark cloth, concealing all but the bulbs. The Uranium bulbs will resemble emeralds; the yellow bulbs, topaz; and the blue, turquoise—certainly a very striking collection of gems. A few diamond-shaped pieces of the foil-coated glass scratched across, by the whiteness of the tiny sparks will aid to set off the whole. The outfit is not expensive: a coil giving a one half inch spark will light from four to six tubes to great brilliancy. Cloths with metallic threads woven in them must not be used, nor any of the metallic powders known in the trade as "glitters."

CHAPTER IX.

GAS LIGHTING BY THE SERIES METHOD.

WHEN it is desired to light clusters of gas jets situated in inaccessible places, or a number of them simultaneously, this method finds ready application. It operates in the division of a long spark among a number of burners, the gas being turned on at the main and the primary circuit of a Ruhmkorff coil closed and opened until the succession of sparks ignites the gas, Fig. 41. There are various commercial forms of these burners, prominent among which is the " Smith jump spark" burner.

A lava tip is provided with a mica or isinglass flange midway between the tip and the lower end of the burner. This flange isolates the electrodes from any possibility of the spark straying away to the metallic pillar in which the burner is inserted. The multiple lava tip burner is intended for use where a very short burner is needed, also for flash rings multiple lights. Here the tips are placed close enough to-

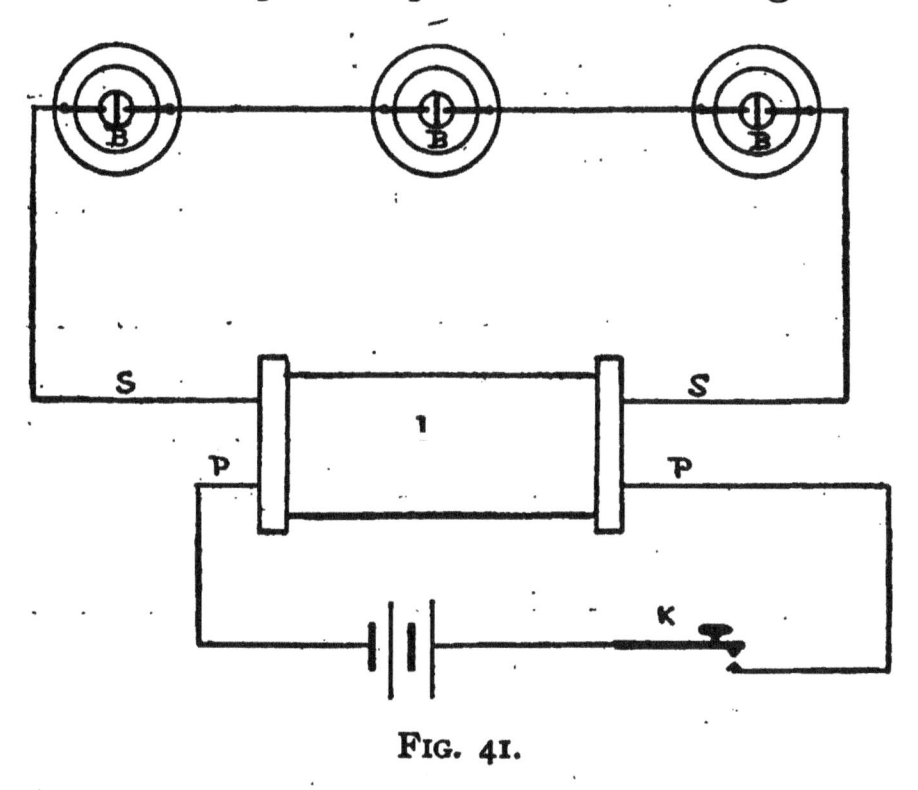

FIG. 41.

gether to ignite by contagion. In this case one of the common tips is removed from the ring and a multiple lava tip substituted. It is customary to allow sixteen burners to one inch of spark. Any number of series can be operated alternately by means of a suitable switch.

The wire used to connect the burners is generally bare copper, and as small in diameter as will sustain its own weight without injury, the amount of the current being infinitesimal. It is supported on porcelain or glass knobs screwed to the wall or ceiling, being carefully planned to avoid any metallic substances to which the spark might be tempted to escape. In wiring chandeliers, the wire is run through glass tubes wherever there is any liability of its coming near the metal pipes. There is a very great danger of this jumping of the spark where it is not wanted, and the utmost care must be taken in planning the course the wires shall take. Even a damp wall will cause trouble or a gilt cornice, although the latter may be entirely insulated from the ground. The switch bases for the groups of circuits must be of hard rubber, and the switch points and levers be placed so far apart that there is no liability of the spark jumping, which it certainly will do if it gets a chance. Ordinary insulated wires are ineffectually protected by the rubber compounds used. Glass, mica, and better still, a large air gap are the only insulations that will serve, for the tremendous potential or voltage of the current must be carefully considered whenever insulation is necessary. The coil is better provided with a spring key in the primary circuit than a vibrator, it gives better control of the circuit and probably a larger and better spark.

OZONE.

An efficient form of apparatus for the generation of ozone is that of Siemens, Fig. 42. It consists of a generator or induction coil, a source of oxygen, a drying bottle, and a reservoir. The oxygen is dried by its passage through sulphuric acid, and passes through the annular space between the inner glass cylinder B and the outer C, where it is subjected to the influence of

118 *Gas Lighting by the Series Method.*

the silent discharge taking place between the opposed glass surfaces. The inner

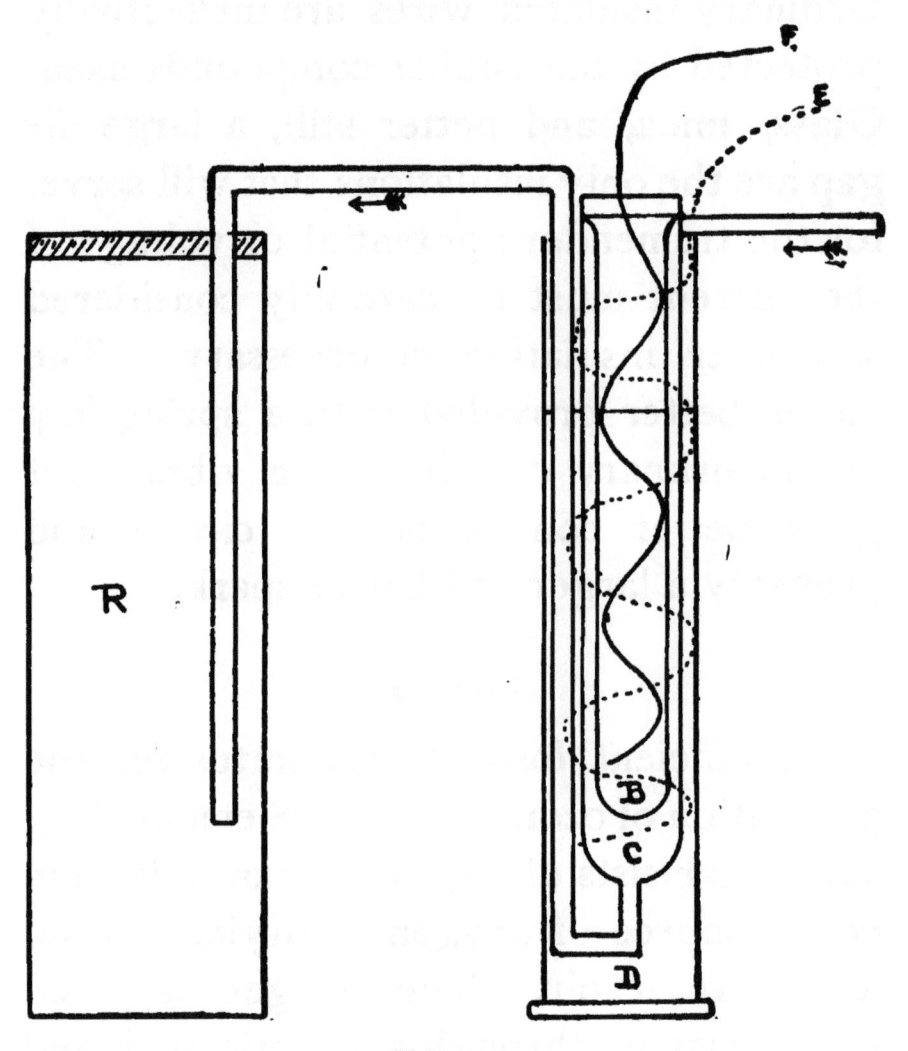

FIG. 42.

cylinder *B* and glass jar *D* are filled with dilute sulphuric acid, and are connected to the coil by the wire solenoids $E\ F$ immersed in the acid. All parts of the apparatus are of glass, the joints being blown or ground in. R is a reservoir to hold the ozone, or the article to be subjected to its influence. Ozone is an active sterilizing agent acting rapidly upon bacilli and pathogenic organisms. When present in large quantities in the atmosphere it acts upon the mucous membranes of the throat and on the lungs, causing intense irritation. It is also used to "age" wines and liquors, and to improve the flavor of coffee; but in the latter case it impairs the appearance of the coffee bean. The bleaching action of ozone in conjunction with chlorine has proved to be of great utility in textile manufactures.

CHAPTER X.

BATTERIES FOR COILS.

IN selecting a battery to operate the coil, one is needed which will supply a large steady current for a considerable period. Although the primary circuit is opened and closed rapidly, yet the class known as open circuit cells is not suitable, even though they have a low internal resistance, and thereby render a large current. Such cells are only suitable for the uses for which they are mostly designed, bell-ringing or annunciator work. There is one case, however, where an open circuit cell may be used with an induction coil, and that is in gas lighting as previously described; but here a dozen or so impulses of current are generally sufficient, followed by long periods of rest. For the latter work the cells in common use are the Samson, Champion, and Monarch, all of which are of low internal resistance and great recuperative power.

The reason that such cells will not work for long periods, is that they polarize. This latter action takes place in these open circuit cells, which are of the Leclanche type as follows: A positive plate of zinc is immersed in a solution of ammonium chloride (or salammoniac), and a negative plate of carbon and peroxide of manganese, contained either in a porous cup or compressed into a block also stands in the solution. Care is taken that these two plates do not touch each other. When the outside circuit is closed the zinc combines with the chlorine of the solution liberating free hydrogen and ammonia. The hydrogen appears at the negative plate, where it is acted upon by the oxygen of the peroxide of manganese to form water.

But when the circuit is of too low resistance, the oxidizing action of the peroxide of manganese is not rapid enough, and a film of hydrogen, which is a poor conductor, forms over the negative plate, increasing the internal resistance of the cell and setting up local action. In the best class of these open circuit cells, this hydrogen is absorbed after a rest, and the battery recuperates and is ready for work again. The circuit of the Ruhmkorff coil is low, and this polarization always occurs a few minutes after the contact-breaker is started.

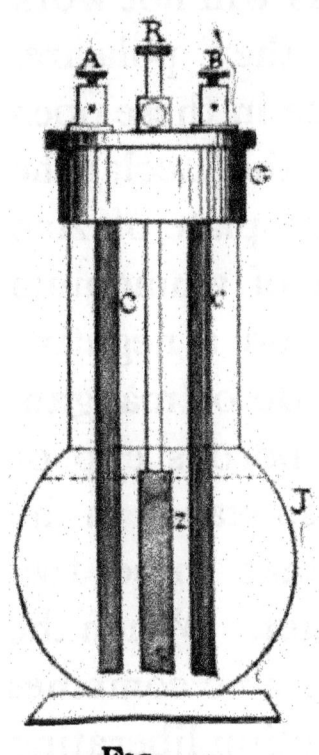

FIG. 43.

In the class of closed circuit cells, chosen for the present purpose, the Grenet or bottle bichromate is one of the handiest for occasional use. A glass bottle-shaped jar, *J*, Fig. 43, is provided with a hard rubber cap, *G*, on which are mounted the binding posts *A B*. To the underside of this cap are attached two carbon plates *C C*, which reach nearly to the bottom of the jar, being connected together on the cap by a varnished copper strip, the latter being in turn connected to one binding post. Through the centre of the cap passes a brass rod, *R*, having attached to its lower end a piece of sheet zinc, *Z*, well amalgamated with mercury. This process of amalgamation consists in cleaning the zinc, then rubbing its surface with a rag dipped in dilute sulphuric acid, and pouring a few drops of mercury on the wet zinc. The mercury will spread readily over the zinc, provided it has been well cleaned, and if properly done should give the zinc plate a bright, shining appearance.

When the cell is not in use, the zinc is drawn up into the neck of the bottle and clamped by a set screw against the brass rod. A copper spring pressing on the rod serves to carry the current to the second binding post.

This cell originated in France, whence its name, but a cheaper form is now made in the United States known as the Novelty Grenet. The shape of the jar is somewhat different, and the carbon is moulded, whereas the French carbon is sawed from the carbon deposited in the gas retort; but the American form is practically of as great utility as the French, and the cost recommends it.

The bichromate solutions are affected by light, and deteriorate less if kept in stoneware jugs. The Grenet battery can very well be fitted into a neat wood case, which will serve the further purpose of preventing chance knocks from fracturing the glass jar.

Carbons which are used in batteries containing the foregoing solution should be well washed in warm water whenever the solution is changed, and especially when it is intended to put the battery out of active service. When the solution acquires a decidedly green hue it should be replaced with fresh. The electromotive force of this cell varies from 1.90 to 2 volts, and the amperage is dependent on the size of the plates, running from 5 amperes upward.

The glass jar is filled up to the commencement of the neck with a solution of bichromate of potash or sodium, called electropoion fluid, and prepared as follows: To 1 gallon of water add 1 pound of bichromate of sodium, mixing in a stoneware vessel. When dissolved add 3 pounds of sulphuric acid in a thin stream, stirring slowly. As the mixture heats on the introduction of the acid, care must be used to pour in the latter slowly. This solution should not be used until quite cold.

The sodium salt is preferable to the potassium, owing to its not forming the crystals of chrome alum, and also on account of its lower cost and greater solubility, the latter being four times greater than that of the potassium salt. The commercial acid used should contain at least

90 per cent pure acid and should be free from impurities. On filling the battery use utmost care not to splash the solution on any of the metal work, or it will cause corrosion. Although the salts in the solution will most likely make a stain, the corrosive action of the acid can be arrested if the solution be splashed on the clothes by the prompt application of ammonia solution.

The "Fuller" cell, Fig. 44, which is another type of the bichromate cell, is one from which a steady current can be obtained for a longer interval than from the Grenet, but the current is less. The electromotive force is the same, but the current is only 3 amperes, except in certain modifications.

In the porous cup is a cone-shaped zinc having a stout copper wire cast in. This

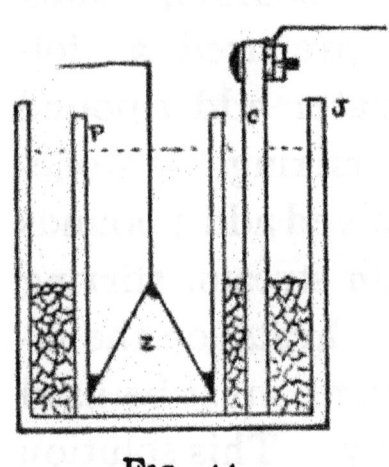

FIG. 44.

wire is occasionally covered with rubber insulation, but, as a rule, is bare. The porous cup is of unglazed porcelain, thick, but very porous. This sets in the glass jar, a wooden cover fitting *loosely* over the whole to exclude dust. Through this cover passes the wire leading from the zinc, and also the carbon plate carrying a machine screw and check nuts for connection. The cover is dipped in melted paraffin, as is also the upper end of the carbon and the rim of the glass jar. This is to prevent the creeping of the salts in the solutions and the corrosion of the brass work.

Into the porous cup is poured a solution composed of 18 parts by weight of common salt and 72 parts by weight of water. Electropoion fluid is held by the glass jar, the two solutions reaching a level of two thirds the height of the jar. One ounce of mercury is added to the porous cup solution to ensure the complete and continuous amalgamation of the zinc. The salt can be more readily dissolved in warm water, but *all* solutions must be used *cold*. It is not always necessary to renew the solutions when the battery fails to give out its accustomed strength, but several ounces of water can be substituted for a similar amount of fluid in the porous cup. Stir the solution by moving the zinc up and down, and a temporary improvement will be noticed.

To obtain a greater current from this cell, use a larger zinc, such as a well-amalgamated zinc plate, and add a teaspoonful of sulphuric acid to clean water for the porous cup solution. Additional carbon plates connected together and placed round the porous cup will lower the resistance of the cell and increase the current, and also tend to keep down the polarization.

A new form of this battery was described by M. Morisot a short time ago.

The positive pole is of retort carbon in the outer cell in a depolarizing mixture made of 1 part sulphuric acid, 3 parts saturated solution bichromate of potash, crystals of the latter salt being suspended in the cell to keep up the saturation. A porous cup contains a solution of caustic soda. The zinc is in a second porous cup placed within the first, which holds a caustic soda solution of greater density. The electromotive force is $2\frac{1}{2}$ volts when the cell is first placed in circuit, and will remain at 2.4 for some hours. The internal resistance is low, but varies with the thickness of the porous cups. This cell is not suitable for any but use for a few hours at one time.

The Dun cell has a negative electrode of a carbon porous cup filled with broken carbon. The zinc is in the form of a heavy ring, and hangs at the top of the solution in the outer jar. Permanganate of potash crystals are placed in the porous cup, and the entire cell filled with a solution of caustic potash 1 part to water 5 parts. The voltage is 1.8, and the internal resistance being low the resultant current is large.

A cell with an electrode of aluminum in a solution of caustic potash and carbon in strong nitric acid in porous cup is claimed to have an electromotive force of 2.8, but the nitric acid is not a desirable acid to handle.

Metallic magnesium in a salammoniac solution with a copper plate in a hydrochloric acid and sulphate of copper mixture is of high voltage, nearly 3 volts being obtained, and the current is large, but it is a new combination and has not as yet stood the test of time.

There are other formulæ for solutions to be used in Fuller or Grenet cells which may be useful to the experimenter. Trouvé's is as follows: Water, 36 parts; bichromate of potash, 3 parts; sulphuric acid, 15 parts, all by weight. Bottone's: Chromic acid, 6 parts; water, 20 parts; chlorate of potassium (increases electromotive force), $\frac{1}{8}$ part; sulphuric acid, $3\frac{1}{2}$

parts, all by weight. A convenient "red salt" or "electric sand": Sulphate of soda, 14 parts; sulphuric acid, 68 parts; bichromate of potash, 29 parts; soda dissolved in heated acid, and potash stirred in slowly. When cold can be broken up and prepared when required by dissolving in five times its weight of water.

The chromic acid used in Bottone's solution is very soluble in water, it being possible to dissolve five or six times the amount in the same quantity of water as of bichromate of potash. The simple solution of chromic acid is 1 pound to 1 pint of water, to which is added 6 ounces of sulphuric acid.

When it becomes necessary to cut zinc plates, it may be readily done by making a deep scratch on the surface, filling the scratch first with dilute sulphuric acid, and then with mercury. The mercury will quickly eat into the metal, and the plate may be easily broken across or cut with a saw. Zinc plates can be bent into shape by the application of heat. Hold the plate in front of a hot fire until it cannot be touched by the bare hand; it will be found that it has softened so that it can be bent around a suitable wooden form. As zinc plates are most attacked at the surface of the acid solution, it is advisable to coat the extreme upper portion of them with varnish or paraffin. Rolled zinc is always preferable to cast, especially so when immersed in acid solutions.

To avoid confusion, it may be stated here that it is the rule to speak of the zinc element as the positive plate and the negative electrode or pole, and the carbon *vice versa*. The portion of the element immersed in the solution is the plate, the part outside, the pole or electrode. In diagrams and also in formulæ positive is shown by a $+$ (plus) sign and negative by a $-$ (minus) sign.

The relation of cost of the materials most used is shown in the subjoined table, which cost, however, varies with the market:

Batteries for Coils.

```
Sulphuric acid, chemically pure................18
      "        "   commercial..................... 1.5
Muriatic      "     ............................ 1.12
Nitric        "     ............................ 3.5
Electropoion fluid............................. 2
Bichromate of potash..........................10.5
      "        "   soda........................ 8.5
Caustic soda.................................. 9
Salammoniac................................... 7
Chromic acid..................................19
Blue vitriol.................................. 4
Litharge...................................... 5.75
Mercury bisulphate............................94
Paraffin...................................... 9
Beeswax...................................35 to 45
Shellac varnish...............................87
Tinfoil.......................................35
```

Gravity Battery.

A cheap modification of the Daniell cell. A glass jar has at the bottom a copper plate consisting of 4 to 6 leaves of thin sheet copper, set on their edges in a star-like shape, a copper wire being attached to the copper rivet which holds the leaves together. A mass of crystals of sulphate of copper is filled in and laid on the top of the copper electrode an inch or so above its top. The negative plate is a variously shaped plate of cast zinc hung from the edge of the jar and reaching about 2 inches from the top into the fluid. Water is poured in until it covers the zinc, and the battery is complete. The sulphate of copper deposits its metallic copper on the copper leaves and liberates sulphuric acid, which rises and attacks the zinc, setting free sulphate of zinc. The sulphate of zinc solution being of greater density remains near the bottom, and the sulphate of zinc solution stays near the zinc. When the cell is left too long on an open circuit the two solutions tend to mix, and copper is deposited on the zinc. The sulphate of zinc finally saturates the top solution, which has to be partly drawn off and replaced by fresh water and crystals of sulphate of copper dropped into the jar to take the place of that which has been decomposed. Electromotive force 1 volt, current from $\frac{3}{10}$ to $\frac{6}{10}$ of an ampere. The

Batteries for Coils. 135

practical working of this cell will be treated of later on in these pages.

The Gethins (Fig. 45) and the Hussey bluestone cells both have the zincs standing in porous cups (shown by dotted lines), which in turn are supported half-way down the jar, generally resting on the copper strip acting as a porous partition between the fluids. The zinc stands in a solution of zinc sulphate, or a weak sulphuric acid solution. The internal resistance is low, and the current large, being from 1 to 5 amperes. These cells are the ideal bluestone cells for charging storage batteries requiring very little attention. The special Gethins cell shown in the figure has the copper made with a collar, which encircles the porous cup, and thereby lowers the internal resistance of the

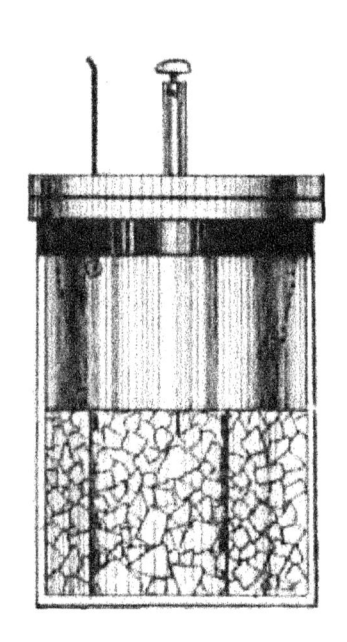

FIG. 45.

battery. The voltage not being over 1 volt, however, renders these cells hardly suitable for direct connection. Five cells connected in multiple would give all of 10 amperes of current, and 1 volt, and a number of these multiple groups could be connected in series for a higher voltage.

CHAPTER XI.

STORAGE OR SECONDARY CELL.

THE development of the storage or secondary cell has been one of the most important electrical advances of the century. For purposes of experiment or work, where a large or steady current is required from compact and readily tended apparatus, the storage cell proves its utility. The simplest form was that used by the early experimenters, and as it is easy to make, a form of it may very well be described.

From a sheet of lead $\frac{1}{8}$ inch thick two or more pieces are cut of the requisite size, say, 5 inches square. In making these plates, they should be cut so as to leave a strip 1 inch wide and 3 inches long, projecting from one corner, *A* (Fig. 46), for the purpose of connection. This is for the reason that the fumes of the sulphuric acid solution would quickly corrode any wires or screws in the plates, and also to give a better connection. The number of plates cut must be an odd one, as it is general to make the two outside plates of

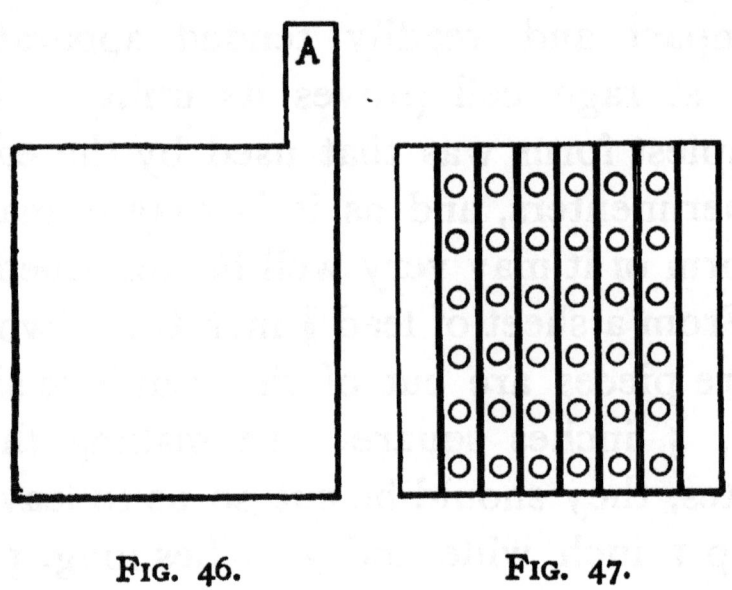

FIG. 46.   FIG. 47.

the same polarity—viz., negative. These plates are then scored with a steel point across and across on both sides to perhaps a depth of $\frac{1}{64}$ of an inch. This scoring is not absolutely necessary; it somewhat hastens the formation of the plates. The plates are then laid face to face, being separated by pieces of wood, rubber, or, still better, by a piece of grooved wood, Fig. 47 having a thin piece of asbestos on each side. These grooves are to carry off the gas, and should run up and down the board, as in the figure. The wood is $\frac{1}{8}$ of an inch thick or thereabouts, and preferably perforated with holes $\frac{1}{4}$ of an inch or larger. When laid together, a few strong rubber bands hold the plates from coming apart. To prevent lateral motion, a few rubber pins may be thrust through the plates. The alternate strips are to be connected together in two series, as in a condenser, and the complete series placed in a jar containing a mixture of seven parts of water to one of sulphuric acid. The terminal of the strips connected to the smallest number of plates is to be marked *P* or +, for positive.

This terminal is now to be connected to a charging current (not over 1 ampere), as described in the directions for charging batteries, for eight hours, and then discharged at a rate not over 1 ampere for six hours. Then the connections are to be reversed and the cell charged backward, as it were, and discharged. This has to be repeated for a long period, perhaps a month, before the cell is in good condition; on the final charge it is to be connected positive to positive of charging source. This operation is called "forming," and the result is to change the metallic lead of the positive plate into red-brown peroxide of lead, and the lead negative plates into spongy lead.

In modern commercial cells this operation is no longer pursued, the plates are variously constructed of lead frameworks holding plugs of litharge or lead oxide, which is "formed" with great facility. For many purposes other than operating Ruhmkorff coils, a few simple cells made, as described, are handy to have and easy to make. In sealing the cells up for portability, care must always be taken to leave a small hole in the cover for the escape of the sulphurous acid gas.

CHARGING STORAGE BATTERIES.

Although the charging of a storage or secondary battery is by no means a difficult operation, yet it requires care, and one unaccustomed to the work will meet many slight difficulties which may seriously affect the results. Pre-eminently the best charging source is a direct current, constant potential electric-light circuit. The amount of current required varies according to the type and make of the cell, but we will select one of a capacity of 50 ampere hours for illustration.

By 50 ampere hours is meant a delivery of 1 ampere per hour for fifty hours, or a rate of discharge equal to the above, as 2 amperes per hour for twenty-five hours. In practice a secondary cell will not be found to act exactly as above, the total amount of current decreasing as the discharge is greater. Each cell is constructed to discharge at a certain rate, above which it is not safe to go. Five amperes per hour is a suitable rate for a fifty-hour cell, and should not be greatly exceeded. The Chloride type, however, is one which can be discharged at a higher rate than normal without any serious results, the latter being generally a bulging or "buckling," as it is called, of the plates whereby they short circuit or fall apart. The voltage of the charging source should be at least 10 per cent over that of the battery when fully charged. The voltage of a cell of storage battery varies from about 2.3 at commencement of discharge to 1.7, at which latter voltage discharge must be stopped and charging recommenced.

Fig. 48 shows the connections to charge a storage battery from an electric-light circuit. The latter must be direct current and of low tension. The circuit from the negative lead runs to the rheostat handle R, thence through as many coils as are in circuit (varied by moving the handle over

Storage or Secondary Cell. 143

the contact pieces in connection with the resistance coils). It leaves the rheostat at *C*, passing to the negative of the storage cell *B*. The positive of the cell is connected to the positive main. The current

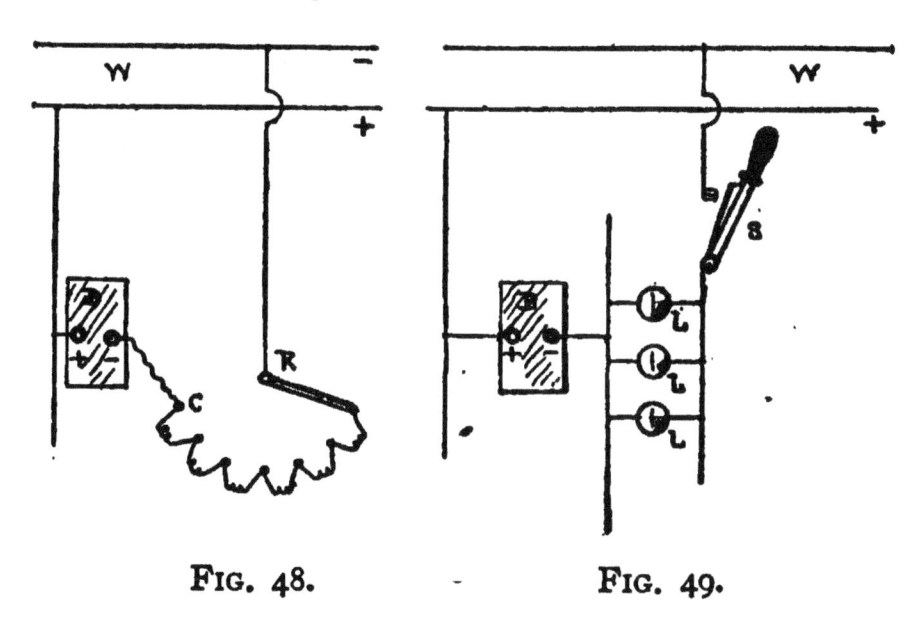

FIG. 48.   -   FIG. 49.

strength is varied at the rheostat according to the charging rate desired.

Fig. 49 shows the employment of lamps instead of the rheostat. The lamps *L L* regulate the current flow by the manner in which the circuit is arranged. If only one lamp be turned on, the current necessary for only one lamp circulates through the battery. Each additional lighted lamp adds to the current by decreasing the resistance of the circuit. *S* is a switch which must always be left open when the dynamos are to be stopped.

Charging from Primary Battery.

In many instances an electric-light circuit is not available for charging purposes, in which event recourse must be made to a primary battery. The one most suited for the work is the modified Daniell, or copper and zinc combination in solutions of sulphate of copper (bluestone) and sulphate of zinc respectively.

There are many good forms of this cell on the market, chief of which are the simple gravity, the Gethins, and the Hussey, which have been previously described. An example will now be described of the operations necessary with the gravity cell, charging one 50-ampere hour storage cell. At least six cells of gravity will be required, as the voltage of each cell is never over 1 volt, and is dependent on the resistance in the external circuit falling as the resistance is lowered. Place the six clean glass jars on a firm foundation, where there is no liability of shaking and no dust likely to settle. Unfold the copper strips into the form of a star, bending the corners for half an inch so as to give an anchorage in the bluestone. Place them into the bottom of the jars and pour in water enough to cover them at least 3 inches below the surface. Now carefully drop in 4 pounds of clean bluestone, which will fill in the angles between the copper wings, at the same time holding the element down to the bottom of the jar. Hang the zincs from the top edge of the jar, and fill up with water to 1 inch from the top. The addition of 5 ounces of sulphate of zinc per cell will render the cells immediately available, and for the further hastening of the chemical action, the copper wire from each copper may be inserted in the binding post-hole of the zinc belonging to its own cell and screwed tight for a few hours; or the cells may be connected together in series, and the wire from the last copper be screwed to the zinc of the first, thus putting the whole series on short circuit. The only advantage of the first method being a saving of time when a number of cells is being set up. This saving of time is often of consequence, as the longer the newly set-up cell is on open circuit, the more copper will be deposited on the zinc, which is highly undesirable. This is shown by the blackening of the zinc as soon as it is put in the solution, which blackening it is hard to prevent entirely. When the cell is working satisfactorily it will show a clearly defined line between the colorless solution above and the deep blue solution beneath.

Gravity cells should never be moved. If no sulphate of zinc is available, half a teaspoonful of sulphuric acid may be poured in over the zinc, which will tend to form the sulphate of zinc. Without

Storage or Secondary Cell. 147

any of these helps the cell will take at least twenty-four hours on a short circuit before it will give its normal current This current should be from $\frac{4}{10}$ to $\frac{5}{10}$ of an ampere. Five cells set up by the writer varied after the addition of the zinc sulphate from 200 milli-amperes (thousandths of an ampere) to 300 milli-amperes, although they were apparently all set up alike ; but after twelve hours' short circuiting they all gave a fairly uniform current of from 470 to 500 milli-amperes.

From time to time on storage battery work, say, every week, the specific gravity of the top solution must be tested with a hydrometer, when the apparatus should be put into the solution and allowed to come to rest. The indicated number at the level of the liquid should be 25°. If the number is higher some solution should be drawn off and clear water added, until

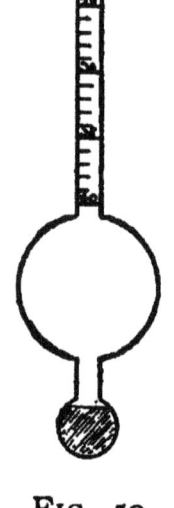

FIG. 50.

the hydrometer settles down to 25° or thereabouts. The inside of the glass jar for 1 inch from the top may be greased to prevent the salts of zinc creeping over the edge, or half an inch of heavy paraffin oil be poured on the top to prevent evaporation and creeping. When the zinc gets very much coated with the dark deposit it must be taken out and scraped and washed. When the bluestone needs replenishing, drop in carefully and be sure none lodges on the zinc element.

SETTING UP THE STORAGE CELL.

Each manufacturer of storage cells issues specific directions for the charging of his own make, but generally the method is as follows : The acid solution is prepared by mixing one volume of sulphuric acid to from four to seven volumes of water, according to the make of the cell. The sulphuric acid should have a specific gravity of 1.82 and be chemically pure. *The acid must always be poured into the water, and slowly, stirring all the time, then set aside for the mixture to cool.* It is best to mix the solution in a separate earthenware vessel, and when two or more cells are to be set up, to mix all the solution at one time, to ensure the same strength, unless a hydrometer is used to determine this.

A good method to ascertain the exact quantity of solution required is to place the elements in the jar and cover 1 inch deep at least with water, then remove the elements and pour off the volume of water corresponding to the proportion of acid to be added, and lastly pouring the remaining water into the mixing vessel, prepare the solution, or electrolyte, as it is called. New elements should be wetted with pure water before being immersed in the solution. An ordinary charge of the electrolyte requires from six to ten hours to cool thoroughly, as considerable heat is evolved in the mixing.

Having now prepared the storage battery solution and set up the primary cells, the charging can be proceeded with. The current must be turned on the storage cell immediately the elements are placed in the acid. Connect the wire from the zinc of the primary battery to the negative of the storage cell and the copper wire to the positive. As the current from a gravity cell is but small, it will take quite a time to charge a storage cell of 50 ampere hours' capacity fully; it is a good scheme to get the cell charged up from a dynamo source, and use the gravity cells to keep it charged; but this cannot always be done, and the gravity battery will do the work in time. As the best storage cells render but 90 per cent of the current put into them, they must be charged over the number of hours for which they are required to deliver current.

When the cell is fully charged the solution will become milky and give off gas freely. This gas in large quantities is detrimental to health, and on no account should a storage cell be *charged* in a sleeping apartment. It affects the throat and lungs, and renders them susceptible to take cold under suitable circumstances. The average voltage of storage cells, when tested with the charging current on, is 2.4 volts, and the lowest they should be allowed to reach is 1.9 volts, unless otherwise specified by the manufacturers.

Cells in poor condition are liable to form a *white* deposit of sulphate of lead, this fault being known as "sulphating." This trouble requires much careful nursing, and the cells must be charged for a long time at a very low rate until the plates of the positive element regain their normal gray color. Chips of straw or excelsior, etc., falling in between the plates will carbonize and cause trouble.

Most portable cells are sealed, but all cells can be easily sealed with paraffin wax for amateur use. Cover the elements fully $\frac{1}{2}$ inch above the normal height of the electrolyte with water before pouring in the electrolyte. Melt some paraffin in an earthenware jar and pour it on top of the water, about the middle of the surface, when it will spread, and care having been taken to have the jar sides dry, will cake solid and form a good seal. Then bore a hole with a brace and bit or some such tool through the wax and pour out the water. The cell can then be set up as usual, the hole being only partly closed to allow of the escape of the generated gas. A glass or rubber tube can be sealed into the hole in the wax, and makes a more finished job.

While on the subject of primary batteries for charging storage cells, a few remarks on their electromotive force may not be amiss. Although the specifications issued by the manufacturers specify an excess charging voltage of 10 per cent over the total voltage of the storage cells, this does not apply to primary cells in its entirety. The voltage of five gravity cells in series would aggregate 5 volts, and the voltage of one storage cell but 2 volts, but

Storage or Secondary Cell. 153

there would not be 5 volts available to force the charging current through the latter. In the first place there is the counter electromotive force of the storage cell working against the gravity battery. Simple subtraction would show only 3 volts excess in favor of the primary electromotive force ; but the working voltage of a galvanic cell varies according to external resistance of the cell and the external resistance of the circuit. When the internal resistance is high, as in the gravity cell, and the circuit resistance is low, in this case being the storage cell, the available electromotive force of the primary is low also.

In many cases it is desired to operate a Ruhmkorff coil from an electric-light main direct. This can readily be done if the circuit be of the constant potential class— that is, one constructed to furnish current for incandescent lamps in multiple. With the direct current, such as the Edison, all that is necessary is either to interpose a rheostat, as in Fig. 48, or to use the lamps, as in Fig. 49. The manner of connecting up is the same as if the storage cell B be replaced by the coil. Using the formula $C = \dfrac{E}{R}$, for example, if the circuit be at 110 volts and the coil require 10 amperes, a resistance of 11 ohms will be required. Or using the lamps in the diagram, Fig. 49, about 20 lamps are to be put in circuit. If the current be an alternating one, the contact-breaker will have to be screwed down or short circuited.

CHAPTER XII.

TESLA AND HERTZ EFFECTS.

THE currents of high frequency used by Tesla in his researches are produced by electrical rather than mechanical means. The alternating current dynamo used by him renders a current of 10,000 alternations per second, but the actual current necessary to the performance of the luminous effects has a frequency of millions of oscillations per second, produced by the discharge of Leyden jars or condensers.

Dr. Oliver J. Lodge, in his "Modern Views of Electricity," shows that the discharge of the Leyden jar is in general oscillatory, the apparently single and momentary spark, when analyzed in a very rapidly rotating mirror, is shown to consist of a series of alternating flashes, rapidly succeeding one another and lasting individually less than one hundred thousandth of a second. The capacity of the condenser and inertia of the circuit regulate the rapidity of these oscillations. A 1 microfarad condenser discharging through a coil of large self-induction, such as one having an iron core, may oscillate only a few hundred times per second. On the other hand, a Leyden jar of the 1 pint size discharging through a short circuit will set up oscillations, perhaps ten million per second; and a still smaller jar would give oscillations away up in the billions. But these small jars are quickly discharged, and require a constant replenishing.

The discharge actually consists of a principal discharge in one direction, and then several reflex actions back and forth, becoming feebler until their cessation. In their vibration they generate waves in the surrounding medium, similar in many respects to sound waves, but of infinitely higher velocity. Their length depends on

Tesla and Hertz Effects. 157

the rate of vibration of the source and their velocity. The microfarad discharge before mentioned will have a wave length of perhaps 1200 miles, the small jar not over 70 feet; and yet the true light wave has only an average length of one fifty thousandth of 1 inch. These waves travel into space until they either die out from exhaustion or are absorbed by some suitable body; but they possess the quality of resonance in a degree like those of sound. Two tuning forks of the same pitch will influence one another—that is, one on being vibrated will start the other in vibration, even at a considerable distance, but the electric waves far surpass them in this respect.

Dr. Hertz made the first practical experiments in this direction with his electric resonator (Fig. 51). This apparatus consisted of a 3-inch spark induction coil, I, the secondary wires $S\,S$ being connected to the copper rods $R\,R$, provided with metal balls $B\,B$, nearly 11 inches in diameter. The discharging balls *D D* were approximated until a satisfactory discharge passed between them. A large wire ring having a spark gap in its circuit was so influenced by the resonance as to show minute sparks passing across this gap even

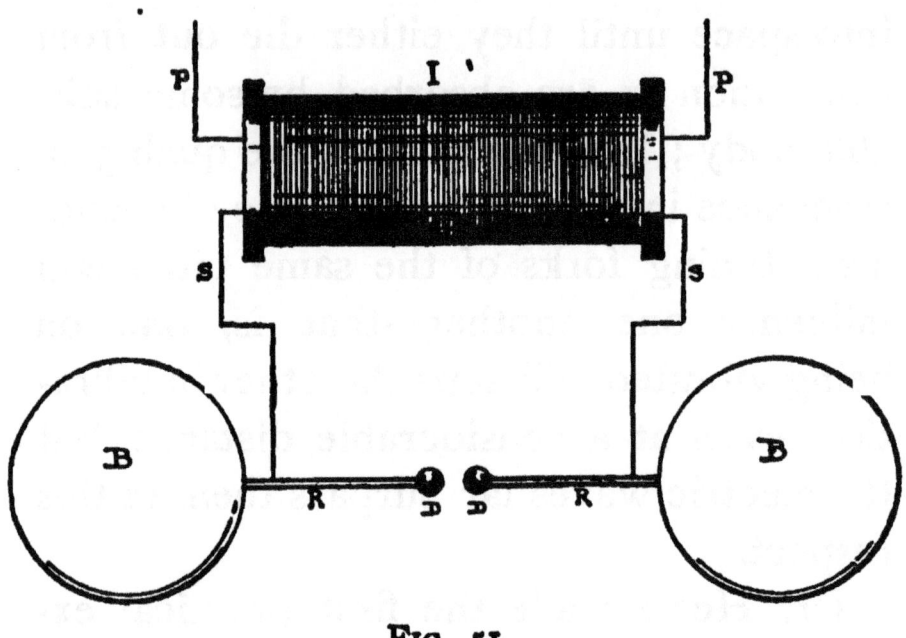

FIG. 51.

when the ring was situated in a distant room. In many experiments with a rapidly vibrating induction coil current, a sparking has been noticed in metallic objects in the same room, in one instance it being discovered in the metallic designs on a wall-paper.

THE "TESLA" EFFECTS.

In exploring the comparatively new field opened up by Professor Crookes, Nikola Tesla has stimulated research into the mysteries of high tension and frequency currents and their effects. In the majority of his experiments Tesla uses alternating currents generated by machinery of his own design, but in a large number of cases his effects can be duplicated with an induction coil suitably energized. In the latter case the apparatus consists of a battery, a Ruhmkorff coil, two condensers, and a second specially constructed induction or disruptive coil, with some few subsidiary implements. The contact-breaker or rheotome must be one giving interruptions of very rapid sequence.

Fig. 52 shows a diagram of the Tesla arrangement with a Ruhmkorff coil. The terminals of the secondary coil of the

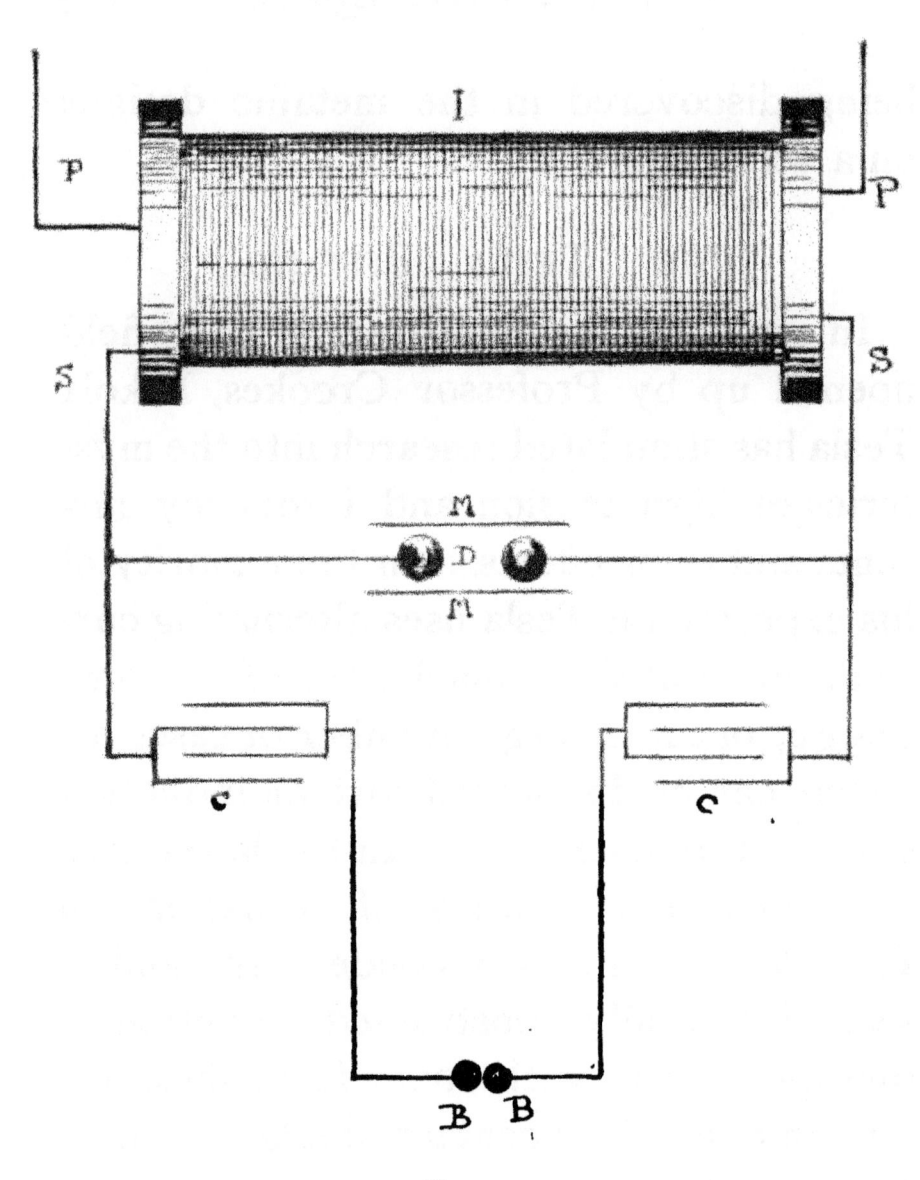

FIG. 52.

Ruhmkorff coil *I* terminate at the condensers *C C*. Bridged across the wires before they reach the condensers is the discharger *D*. The second terminals of the condensers are led through the split primary of the disruptive coil, terminating at the points *B B* of the second discharger. The secondary of the disruptive coil is either outside or inside the primary coil. The condensers are of special design, being small, but of high insulation. They each consist of two plates of metal a few inches square immersed in oil and arranged so they can be brought nearer together or further apart, as necessary. Within limits, the smaller these plates are the more frequent will be the oscillations of their discharge. They also fulfil another purpose, they help nullify the high self-induction of the secondary coil by adding capacity to it.

The discharger tips are preferably metal balls under 1 inch in diameter. Tesla uses various forms of dischargers, but for experimental purposes the two metal balls will answer. They are adjusted when the whole apparatus is working according to the results desired. The mica plates serve

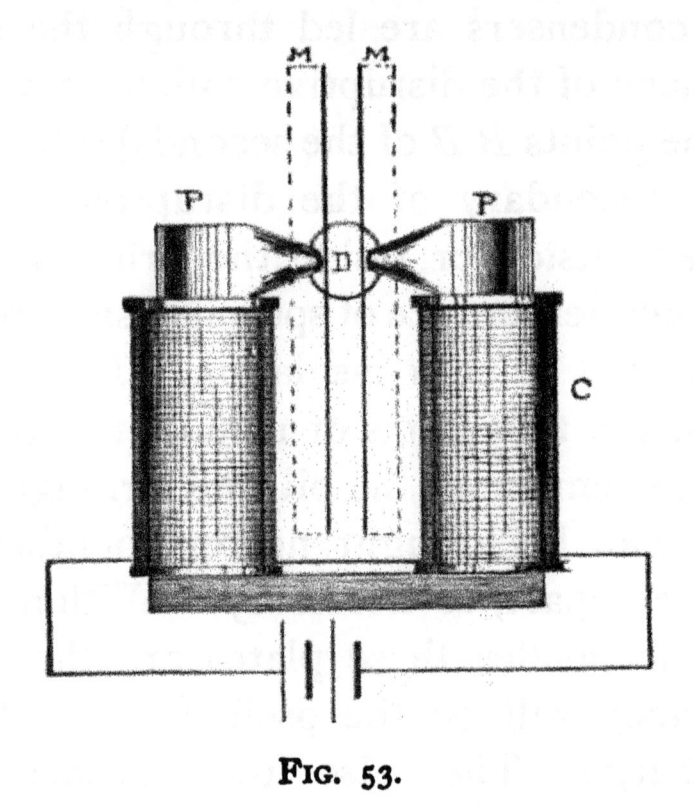

FIG. 53.

to establish an air current up through the gap, making the discharge more abrupt, an air blast being also used at times for the furtherance of this object. The device (Fig. 53) consists of an electro-magnet, *C*, set with its poles P across the air gap, helping to wipe out the spark, as in a well-known form of lightning arrester. This form, described by Tesla, has the pole

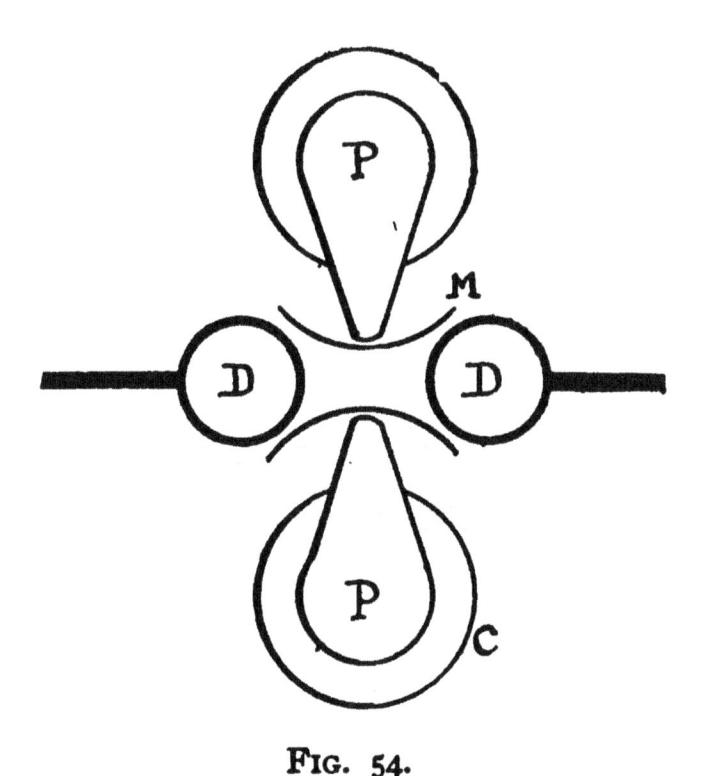

FIG. 54.

pieces P shielded by mica plates M, to prevent the sparks jumping into the magnets. Fig. 53 is an elevation and Fig. 54 a plan of this device. The terminals from the condenser are led to the primary of the

164   *Tesla and Hertz Effects.*

disruptive coil, or, as in Fig. 52, to two discharger balls *B B*. The disruptive coil (Fig. 55) is easily made, and will give good results with careful handling.

The secondary coil *S* of 300 turns of No.

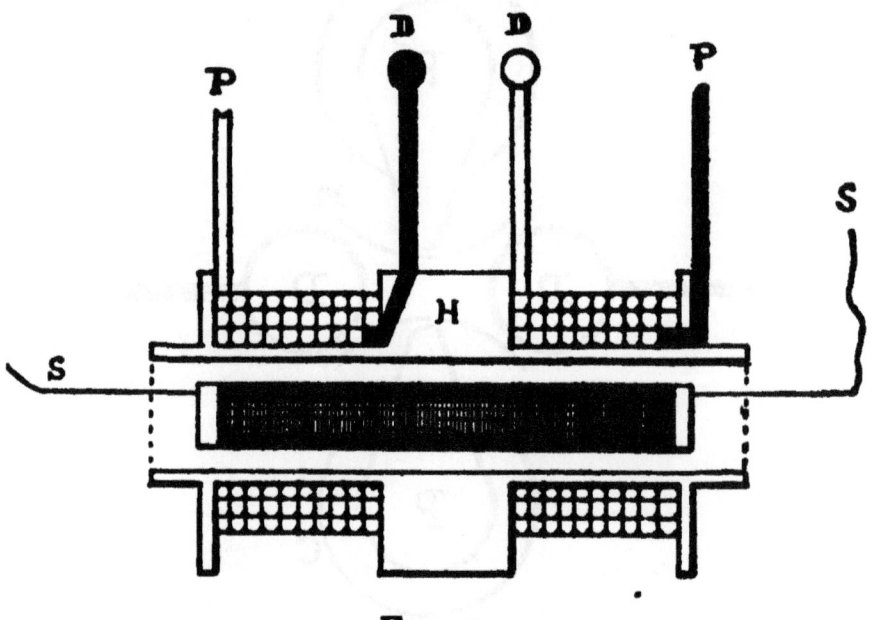

FIG. 55.

30 silk-covered magnet wire, wound on a rubber tube or rod $\frac{1}{2}$ inch in diameter, with the ends brought out in glass or rubber tubes. The primary coils are wound on a second rubber tube, *H*, at least $\frac{1}{16}$ inch thick, and large enough inside to slip easily

over the secondary coil S. It must be long enough to project 2 or 3 inches over the ends of the secondary coil turns. The primary has 50 turns in each coil of No. 18 B. & S. gutta-percha covered wire. The four ends, *P P P P*, of this coil are brought out in rubber or glass tubes, two to the condenser and two to the discharger *D D*. Each layer of these coils must be separated by rubber cloth or even cotton cloth. The further apart the layers the less the inductive effect, but the better the insulation. The whole coil is immersed in a glass or wooden vessel containing boiled out linseed or petroleum oil. A perfect coil should have all the air bubbles drawn out of it after being immersed by putting it under an air-pump, but for many experimental uses this is not absolutely necessary; but extreme care must be taken to insulate the coils and their terminals. This will be apparent when the current is turned on.

The resonance effects obtained during

the operation of a Tesla coil are very marked, and their study may lead to the solution of the problems of communication between distant points without the use of other conducting media than the atmosphere. But the main use to which the Tesla currents have been put is that of artificial illumination. These currents have enabled experimenters to obtain a high luminosity in vacua by the aid of only one conducting wire—in fact, in some cases without the utilization of any conductor than the air. An ordinary incandescent lamp connected to one terminal of the coil will show in a fair degree some of the luminescent phenomena. The brush effects from the terminals of the secondary coil are extremely marked and interesting; but to detail the experiments that can be performed with the Tesla disruptive coil would be an impossibility here. Reference is recommended to the published works of Nikola Tesla, which happily are readily procurable.

These currents of high frequency have of late been turned to account in electrotherapeutics, principally for the stimulation they exert on the nutritive process. They also exert a very great influence on the vasomotor centres, as is evidenced by the reddening of the skin and exudation of perspiration. This result is readily obtainable by placing the patient in connection with one electrode on an insulating stool, and terminating the other electrode at a large metal plate situated a few feet distant ; or the patient may be surrounded by a coil of wire in connection with the coil of sufficient diameter, however, to prevent contact.

CHAPTER XIII.

THE "ROENTGEN" RAYS AND RADIOGRAPHY.

ALTHOUGH the remarkable discovery that it was possible by electrical means to depict an image of an object on a photographic sensitized plate, despite the intervention of solid bodies, was first given to the world at large by Professor Roentgen, yet he was undoubtedly led to the results by consideration of the works of previous experimenters in electrical discharges through vacua.

It is not intended here to trace the previous work of Professor Crookes, the inventor of the radiometer, which is actuated by the heat rays of light, nor of Hertz, who found that gold leaf was transparent to rays emanating from certain vacuum tubes carrying a luminous electrical discharge. Nor can we stop to investigate the experiments of Philip Lenard, who replaced a portion of the tube by an aluminium window, and yet obtained fluorescence from the rays passing through this window.

Professor Crookes claimed to have discovered a fourth state of matter—viz., "radiant matter," the previous three states being, as every student knows, gaseous, liquid, and solid. This he inferred from the effects obtained by the passage of high-tension currents through glass tubes which he exhausted of air to a far greater degree of attenuation than previously attained. Certain ones of these tubes have been mentioned in a previous chapter and illustrated in Figs. 31, 33, and 34. But these earlier experimenters attributed their results to the action of the simple cathode rays (rays emanating from the tube terminal termed the cathode from its connection to the negative pole of the generator).

They had not as yet become cognizant of certain rays now known as Roentgen or X rays, which were to be found also at the cathode.

In a previous chapter, when dealing with the spectrum, mention was made of the ultra violet rays which possessed so high a rate of vibration. When Professor Roentgen first made known the fact that he had succeeded in photographing the bones of a human hand through the flesh by means of a vacuum tube, many scientists claimed the results were obtained from these ultra violet rays. But from the researches of W. M. Stine and others it has been shown that this is not absolutely true. Mr. Stine, by using a very rapid plate (Seeds 27 X) and a slow Carbutt B 16 plate, was unable to detect any difference in the results from the " light speed" of the plates, which light speed is determined by reference to the ultra violet rays.

The connections of the apparatus neces-

FIG. 56.

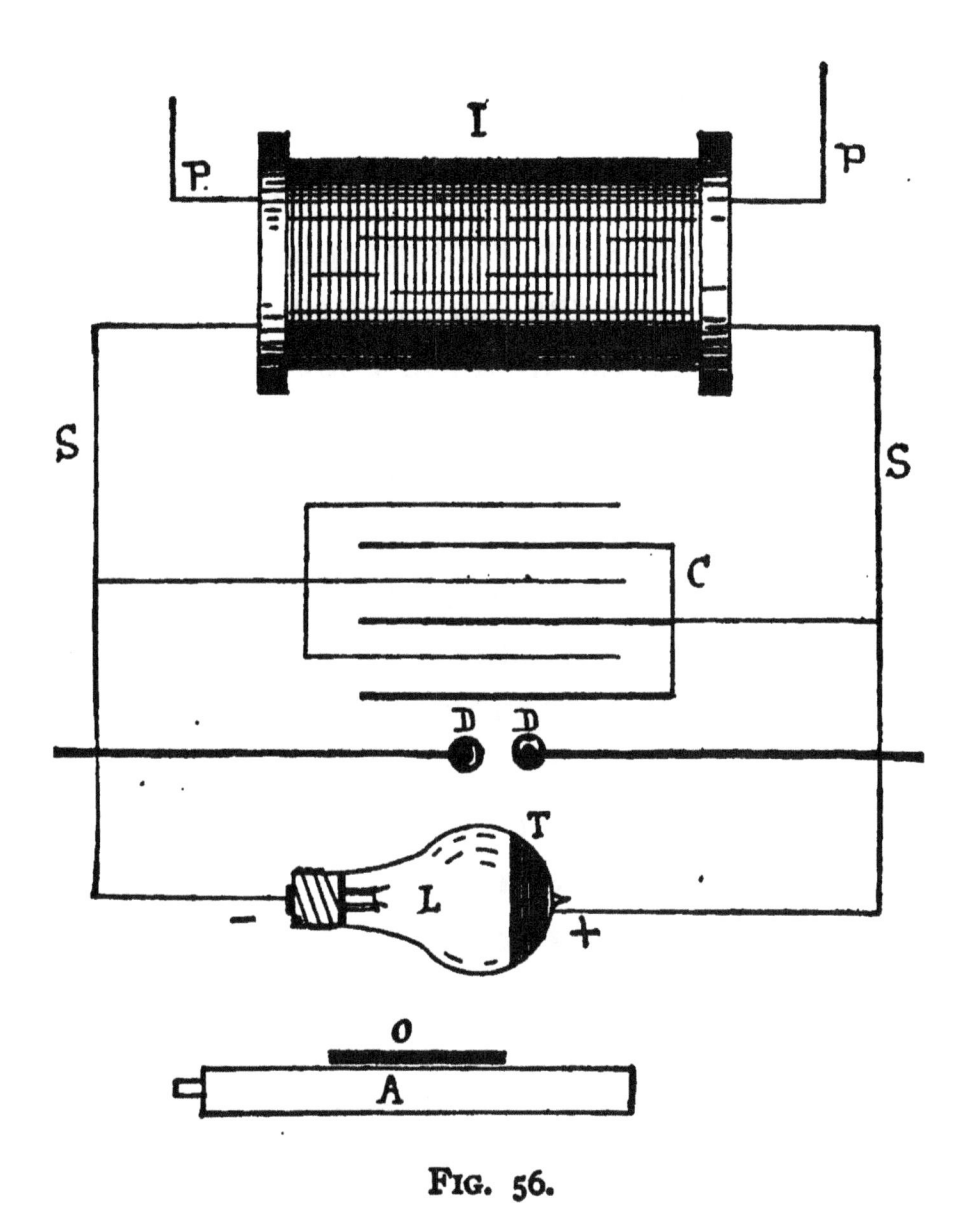

sary for the production of these photographic images or radiographs is shown in Fig. 56. I is the Ruhmkorff coil, C the condenser, L an incandescent lamp or a Crookes tube, $D\ D$ secondary discharger for disruption, O the object to be depicted on the plate contained inside the closed plate-holder A. Connection is made to the lamp by its socket and by a strip or cap of tinfoil cemented over the end of the bulb. In many of the arrangements of the apparatus a lead plate, pierced with a central aperture about 1 inch in diameter, is placed between the tube or lamp to act as does the diaphragm of a photographic lens.

The best results have been obtained from variously constructed Crookes tubes, the earlier design of which is shown in Fig. 57. A later form is that of a cylinder having a bulb at one end, into which, near its junction to the tube, is sealed a second glass tube having an electrode projecting into it. The upper portion of the main

"*Roentgen*" *Rays and Radiography.* 173

cylinder carries a disk-shaped electrode. These two electrodes are so arranged as to focus the rays through the bulb on to the object and sensitive plate.

Elihu Thomson proposes a standard form of bulb-shaped tube having opposed to each other two concave aluminum electrodes attached to wires sealed in the glass. These electrodes focus upon a V-shaped piece of platinum

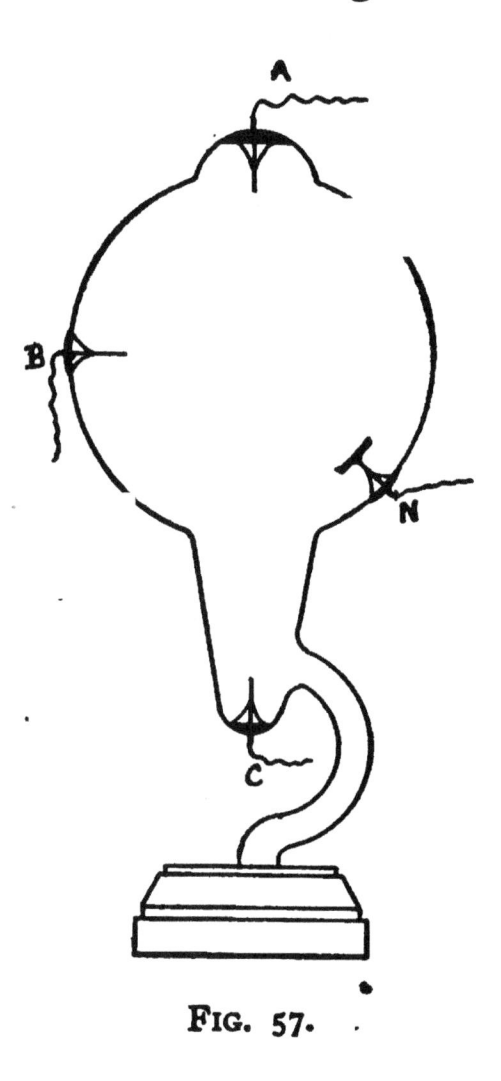

FIG. 57.

or other suitable metal, which in turn reflects the rays downward through the glass side of the bulb. These concave electrodes being connected, as in Fig. 56, become in turn cathodes as the current alternates.

In working with these tubes or lamps no metallic objects should touch the glass, otherwise a deflection of the rays will ensue. Tesla obtained radiographs when the object and the plate were many feet apart, but, generally speaking, the object and the plate should be in proximity.

A number of experimenters obtained images of objects laid on the outside of the plate-holder, as in the figure, by exposure to the rays of the sun; but these have been shown conclusively to be due entirely to light filtering through the plate-holder slide. Professor Elihu Thomson was for a short time of the opinion that the X rays proceeded from the anode or terminal connected to the positive pole of the coil, but more extended research convinced him of their cathodic origin. He also found that fluorescence played no helping part in these effects, since non-fluorescent substances give out the X rays, but was even detrimental to the rays. Mr. Stine found that the rays had their source at that portion of the bulb of the tube opposite the cathode terminal, but had not determined whether the interior or exterior of the tube is active. Professor Roentgen stated that when the radiant matter was deflected by means of a magnet (see Gordon's "Electricity and Magnetism," or reports of Crookes' researches) the X rays proceeded from the spot to which the former was drawn. C. E. Scribner and F. R. McBerty, in a series of experiments with a lead plate pierced with a hole through which the rays passed, confirmed this statement, and also found that the rays proceeded from the inner wall of the tube at the point of deflection of the cathode rays. Professor Pupin has found that internal electrodes are not needed in the tubes, having obtained perfect results from metallic bands on the external surface of the tubes (at each end of the tube).

As yet no place in the spectrum can be definitely assigned to the X rays, it not being proven whether they belong in the visible or invisible portions.

Although images can be produced on sensitive plates when the object is laid on the plate, which in turn is held between two metal plates connected to the coil terminals, yet this is not necessarily radiography from the X rays. The metal plates are put sufficiently far apart to permit of no visible luminous discharge taking place. It is probable that this effect is owing to the same causes as produce an image (visible on being breathed upon) on the surface of a glass plate from a coin laid thereon, and into which the discharge of a Holtz or similar machine is passed. For radiographs, a moderately rapid plate appears to give the best results.

That glass is not absolutely necessary for the tubes is proven by the excellent results given from the employment of exhausted aluminium tubes. Edison, however, used the thin glass bulbs similar to those of the incandescent lamp, and found an improvement over the use of thicker glass. He also found that the duration of exposure of plates should be nearly proportional to the square of the distance from the source of the rays. When his tube was one quarter of an inch from the object laid on the vulcanized fibre slide of the plate-holder, he obtained a good radiograph in less than one second. At 2 feet the time necessary became nearly three minutes, and at 3 feet distance the time increased to over seven minutes.

The opacity of certain substances to X rays, and therefore their degree of suitability for radiography, varies. Their order, as ascertained by Mr. Edison, was as follows: Steel was practically impenetrable; next in order were zinc, lead, glass, and aluminium, while rubber and celluloid can hardly be "radiographed." Professor Terry, of Annapolis, exhibited in the *Electrical Engineer* (New York, April 8th, 1896) a remarkable radiograph of a large number of substances. Some of the most penetrable substances on the print were aluminum, wood, paraffin, mica, and carbon. Lead has been found the most opaque metal in the majority of experiments. Solutions of bromide of potassium, bichromate of potash, and chloride of antimony are rather opaque, but permanganate of potash solution less so.

Mr. Edison has devised an extremely simple apparatus called the Fluoroscope, by the use of which it is possible to see the results with the naked eye without resort to the complicated photographic process. It consists of a pasteboard cone-shaped box, the smaller end being arranged to fit over the eyes, and the larger end closed by a paper screen coated on the inside with tungstate of calcium. This preparation was found to be very much more suitable than the barium platino-cyanide used in previous experiments. By placing the object to be examined between the source of the X rays and the screen, the shadow of the object is thrown on the screen in such manner as to be discernible to the observer looking into the tube. From late evidence, it would seem that Professor Pupin was the first to use apparatus similar to the Fluoroscope. Whether this is so or not, it is certain that Professor Pupin has made many important improvements in the radiographic field. He found that most new tubes required electrical nursing before they worked at their best —that is, they should be alternately excited and cooled for a period determined by actual experiment.

Professor Salvioni, of Rome, Italy, devised an apparatus which he named the Cryptoscope, which bears a strong resemblance to the Fluoroscope, excepting that it is provided with a lens at the eyepiece, and is not rendered sensitive by the tungstate of calcium, as in the latter instrument.

To conclude, the investigation of the

X ray phenomena will undoubtedly lead to many useful discoveries. The making of radiographs is not difficult, provided the experimenter has a good general knowledge of the operation of the coil and of elementary photography. At the present time, when it is but a new discovery, radiography has developed new fields for research, which will amply repay the explorer who thoughtfully and conscientiously pursues his way through them; but, above all, let him hesitate twice before announcing as decisive any new result, which in its novel fascinating plausibility seems to overthrow the old-established theories with which it apparently conflicts.

INDEX.

A.

Acid, Chromic, 131.
Acid, Sulphuric, 148.
Air pump, Geissler, 92.
" " Simple, 91.
" " Sprengel, 93.
" blast, 45.
Amalgamation, 123.
Assembly of coils, 22.
Attraction, Window, 104, 113.

B.

Ballistic galvanometer, 54.
Base, coil, 20.
Battery, Bichromate, 122.
" Champion, 121.
" Daniell, 133.
" Dun, 129.
" Fuller, 126.
" Gas-lighting, 121.
" Gethins, 135.
" Gravity, 133.
" Grenet, 122.
" Monarch, 121.
" Morisot, 130.
" Novelty, 124.
" Open circuit, 120.
" Polarization, 121.
" Samson, 121.

Battery, Solutions, 127, 130, 131.
" Storage, 137.
" Storage, to charge, 141, 144.
" Storage, to make, 138.
" Storage, to seal, 151.
Beeswax, 50.
Brush, Electric, 78.
Burner, Gas-lighting, 114.

C.

Capacity of condenser, 55.
Carbons for battery, 124.
Cathode, 169.
Cements, 52.
Charging condenser, 65.
Chromic acid, 131.
Closed magnetic circuit, 6.
Coil, Disruptive, 164.
" Primary, 7.
" Secondary, 10.
" Winding, 20.
Condenser, 28, 54.
" Charging, 65.
" Capacity of, 55.
" Discharge of, 156.
" Glass, 56.

Index.

Condenser, Mica, 63.
" Paper, 60, 62.
 Series, 63.
Contact breaker, 26.
" " High speed, 34.
" " Polechanging, 42.
Core, 4.
" Length of, 7.
Cryptoscope, 179.
Current, Extra, 1.
" Physiological effect, 167.

D.

Dielectric, 59.
Discharger, 26.
Dun cell, 129.

E.

Eddy currents, 6.
Electric sand, 131.
Electrode, 132.
Electrolyte, 149.
Ends for coil, 25.
Extra current, 3.

F.

Farad, 55.
Fluoroscope, 178.
Fluorescence, 87.
Foucault currents, 6.
Fraunhofer's lines, 85.

G.

Galvanometer, 54.
Gas lighting, 114.
Gassiot star, 103.

Geissler tubes, 109.
Glass, To pierce, 80.

H.

Hertz resonator, 157.
Hydrometer, 147.
Hysteresis, 6.

I.

Induction, 1.
" Self, 8.
Insulating cements, 52.

L.

Leyden jar, 54.
Lighting gas, 114.

M.

Magnetic circuit, Closed, 6.
Mercury contact breaker, 40.
Mica condenser, 61.

O.

Oil, Linseed, 48.
" Mineral, 47.
" Resin, 51.
" Spark through, 50.
Opacity of substances, 177.
Ozone, 71.
" Apparatus for, 117.
" Uses of, 119.

P.

Paper condenser, 62.
Paraffin, 49.
Polarization, 121.
Pole, 132.
Polechanging switch, 32.
" contact breaker, 42.
Primary coil, 7.

Index.

R.

Radiant matter, 169.
Rays, Ultra-violet, 170.
Reel ends, 25.
Resonance, 157.
Resonator, 157.
Resin oil, 51.
Rheotome, 2.
Roentgen, 168.
Rotating wheel, 104.

S.

Salvioni, Cryptoscope, 179.
Shellac, 51.
Siemens, Ozone apparatus, 117.
Signs, Battery, 132.
Soda, Bichromate of, 125.
Spark, Electric, 70.
Spectroscope, 82.
Spectrum, Solar, 82.

Sulphating, 151.
Switch, Polechanging, 32.

T.

Table of cost, 133.
" " wire, 33.
Tesla, Currents of, 155.
Transformer, 5.
Tube, Insulating, 9.
Tungstate of calcium, 178.

W.

Wax, 50.
Wheel, Rotating, 104.
Winder, Coil, 16.
Winding coils, 20.
" Sectional, 11.
" Secondary, 10, 14.
Wire for secondary coil, 24.
" " primary coil, 9.
" Sizes and weights, 33.

BOOKS ON ELECTRICITY

Telephones, their Construction and Fitting. A Practical Treatise on the Fitting Up and Maintenance of Telephones and all the necessary Apparatus required. By F. C. ALLSOP. 256 pages, 208 illustrations and folding plates. 12mo, cloth, $2.00.

Electric Bell Construction. A Treatise on the Construction of Electric Bells, Indicators, Electro Magnets, Batteries, Relays, Galvanometers, etc. 131 pages, 177 illustrations. 12mo, cloth, $1.25.

Practical Electric Bell Fitting. A Treatise on the Fitting Up and Maintenance of Electric Bells and all the necessary Apparatus. By F. C. ALLSOP. 170 pages, 186 illustrations and folding plates. 12mo, cloth, $1.25.

Induction Coils, and Coil Making. A Treatise on the Construction and Working of Shock, Medical and Spark Coils. By F. C. ALLSOP. 162 pages, 118 illustrations. 12mo, cloth, $1.25.

Wrinkles in Electric Lighting. By VINCENT STEPHEN. 45 pages, 23 illustrations. 12mo, cloth, $1.00.

Short Lectures to Electrical Artisans. Being a Course of Experimental Lectures delivered to a Practical Audience by J. A. FLEMING, M.A., D.Sc. 208 pages, 73 illustrations. 12mo, cloth, $1.50.

Practical Electrical Notes and Definitions. For the Use of Engineering Students, and Practical Men. By W. PERREN MAYCOCK. 256 pages, 130 illustrations. 32mo, cloth, 75 cts.

Electrical Tables and Memoranda. By Prof. S. P. THOMPSON. 128 pages, illustrated. Vest pocket edition, gilt edges, 50 cts.

Notes on Design of Small Dynamos. By GEORGE HALLIDAY. 79 pages, illustrated, and a number of drawings to scale. 8vo, cloth, $1.00.

LIST OF BOOKS.

All books in cloth unless otherwise stated. Sizes: **F** *folio,* **Q** *4to* **O** *8vo,* **D** *12mo,* **S** *16mo,* **T** *24mo,* **Tt** *32mo,* **Fe** *48mo,* **Ss** *64mo.*

*Adams, H. Designing iron work, new ed., plates, O,	3.50
—— Handbook for mechanical engineers, D,	2.50
—— Hydraulic machines past and present, O, paper,	.40
—— Joints in woodwork, plate, O, paper,	.40
—— Strains in ironwork, plates, D,	1.75
Allen, J. R. Design and construction of dock walls, Q,	2.50
Allsop, F. C. Electric Bells. D,	1.25
*—— Electric bell construction. D,	1.25
—— Induction coils, D,	1.25
—— Telephones: their construction and fitting. D,	2.00
—— R. G. The Turkish bath, O,	2.50
—— Public Baths and Wash-houses, D,	2.50
—— The hydropathic establishment, O.	2.00
*Anderson, R. Lightning conductors, third ed., O,	5.00
Andre, G. G. Coal mining. 2 vols, plates, Q,	15.00
—— Mining Machinery. 2 vols., plates, Q,	15.00
—— Plan and map drawing, illus., Q,	3.75
—— Rock blasting, plates, O,	3.00
Andrews. Life of Railway Axles, O, paper,	.80
*Applebys Handbooks: Mining Machinery, O,	1.00
*—— Machine and Hand Tools, O,	1.50
* Atkins, G. W. Water Softening and Filtration, O,	.40
Babbage, C. Tables of logarithms. O.	3.00
—— Calculating machines. plates, Q,	8.50
Baker, B. Long and short span railway bridges, D,	2.00
*——, T. Formulæ, Rules and Tables for Engineers, D,	.40
Bales, T. The builder's clerk, D,	.60
Barber, T. W. Engineers sketch-book. O,	3.00
*—— Repair and Maintenance of Machinery, O,	3.50
Barlow's Tables of squares, cubes, etc., D,	2.50
—— C. The new Tay bridge. F, illustrated	8.50
Barnaby, W. S. Marine Propeller, O.	4.25
¶ Battershall, J. P. Food adulteration and detection, O,	3.50
Bayley, T. Pocket-book for chemists, 5th ed., Tt. roan *press*	
Binns, W. Orthographic projection, 11th ed., illus., O,	3.50
—— Second course of orthographic projection, O,	4.50
*Bjorling, P. R. Handbook on pump construction, D,	1.50
—— Construction of Pump Details, D.	3.00
—— Direct-acting pumping engine. D,	2.00
Bjorling, P. R. Pumps: historically treated, D,	2.50
—— Water or Hydraulic Motors, illus., D.	3.50
*—— Pumps and Pump Motors (12 parts), pap., Q, each	1.50
*—— Complete in paper $15.00; in 2 vols. half-moroc.	20.00
Boulnois, H. P. Hints on taking a house, S,	.50
—— Munic. and san. engineers' hand-book, O,	6.00
—— Dirty dustbins and sloppy streets, D,	.40
Bow, R. H. Bridges and Roofs, plates, O,	2.00

Box, T Mill-gearing, wheels, shafts, etc., D,	3.00
—— Practical treatise on heat. 5th ed., D,	5.00
—— Strength of materials, plates, O,	7.25
Box, T. Practical hydraulics, plates, D,	2.00
Britton, T. A. Dry rot in timber, plates, O,	3.00
*Britten, F. J. Watch and Clockmakers' Handbook,	2.00
*—— Former Clock and Watchmakers and their Work, D,	2.00
Brook, J. French measures and English equivalents, T,	.40
*Brooks, C. P. Production of cotton cloth, illus, D,	2.25
*—— Weaving calculations. D,	2.00
Broom, G. J. C. Drainage with regard to health 8vo, pap.	.80
Brown, H. T. 507 Mechanical movements, illus., D,	1.00
*Bryers, T. Assistant to Practical Cotton Spinning, D,	1.00
*Buchanan, E. E. Tables of squares, S.	2.00
*Buckley, R. B. Irrigation Works in India and Egypt, plates and maps, Q.	25.00
*Burns, W. Illuminating and heating gas, D,	1.50
*Burns, G. J. Glossary of terms used in architecture, S,	1.40
‡Busbridge, W. Arch. drawing copies, 38 sheets, each	.10
‡—— Engineering drawing copies. 66 sheets, each	.10
‖ Bury's Marine Propellers, Q,	5.00
‡Byrne, O. Practical mechanics, illus., D,	2.00
—— Method of solving equations, O, paper,	.40
Calkins' Steam engine indicator. O,	1.50
Campin, F. Practice of hand-turning, 3rd ed., D,	2.00
‡Carpenter, W. L. Manufacture of Soap, Candles, D,	4.00
Charleton, A. G. Tin mining, illus., O,	4.25
Christopher, S. Cleaning and scouring, T, paper	.20
Clarke, G. S. Perspective explained, plate, O,	1.25
—— Practical geometry. 2 vols., plates, Q and O,	4.00
—— Principles of graphic statics, plates, Q,	5.00
* Clark, L Transit instrument, O,	2.00
*—— Dictionary of metric measures. D,	2.50
Codrington, T. Maintenance of macadamised roads, O,	3.00
Cole, W. H. Permanent-way material, O,	2.25
Cole, T. Institute of Gas Engineers, Vols. 2-4., D. each	8.40
Collet, H. Water Softening and Purification, D,	2 00
Colyer, F. Apparatus and fittings for gas works, O,	5.00
—— Hydraulic machines. 2nd ed., plates, O,	10.00
Colyer, F. Management of steam boilers and engines, D,	1.50
—— Modern steam engines and boilers, plates, Q,	5.00
*—— Modern Sanitary Appliances, D.	2.00
—— Public institutions. O,	4.25
—— Pumping Machinery, 2nd ed., plates, v. 1, O.	10.00
—— Pumps, and pumping machinery, plates, v 2., O,	5.00
*—— Water supply, drainage, and sanitary appliances, D,	1.50
‖ Cooke, Sir F. W. Private Letters and Memoir of O,	1.25
Corliss Engine, Supplement—Slide and piston-valve geared steam engines, 2 vols., plates, F, hf. mor.,	14.00
Corliss Engine and allied steam motors, 2 v., hf. mor, F,	21.00

Corliss Engines. Their management, illus., S, 1.00
*Cornes, C. Mining Machinery, D, 1.00
‖Cotterill, J. The steam engine, second edition. O, 4.50
‡Cousins, R. H. Strength of beams and columns, O, 5.00
Cromwell, J. H. Easy lettering, Q, .50
Cross, C. F. and Bevan, E. J. Paper-making, illus., D 4.00
*Cross, Bevan, and King. Report on fibrous substances, 2.00
Cullen, W. Construction of waterwheels, plates, O, 2.00
*Cunningham, D., Earthwork Tables, D, 4.25
Cutler, H. A. and Edge, F. J. Setting out curves, Tt, 1.00
Dahlstrom, K. P. The firemans guide, 5th ed., D, .50
Davey, H. Differential expansive pumping engine, O, .80
Davies, P. J. Standard practical plumbing, Q, 3.00
‡—— Ditto, Vol. 2, Q, 4.50
Dearlove, A. Working speed of cables. Tt, .80
*Delano, W. H. Natural Asphalt and Bitumen, D, paper, .50
Denning, D. Woodcarving for amateurs, illus., pap., D, .40
‡Denton, J. B. Agricultural drainage, O, paper 1.00
*—— Intermittent downward filtration, O, 2.00
Diesel, R. Rational Heat Motors, O, 2.50
Dixon, T. Millwright's guide, 6th ed., D, 1.25
Donaldson, W. Constructing oblique arches, O, plates, 1.50
—— Solid beams and girders, O, 1.50
—— Tables for platelayers, plates, D, 1.50
—— Water wheels, O, 2.00
—— Transmission of Power by fluid pressure. O, 2.25
Drysdale, J., and Hayward, J. W. Housebuilding, O, 3.00
Dubelle, G. H. Soda fountain drinks, D, 2.50
Du Moncel, Th. Electro-motors, tr. by C. J. Wharton, D, 3.00
*Dunbar, J. Practical papermaker, 3rd edition, T, 1.00
Dye, F. Fitting hot-water apparatus, illus, D, 1.00
—— Hot water fitting and steam cooking apparatus, S, .50
—— Popular Engineering, Q, 3.00
*Ede, G. Management of steel. 5th ed., D, 2.00
*—— Gun material, S, 2.00
Electrics, (Practical.) A universal handy-book, D, .75
*Eldridge, J. Fixing hot-water apparatus, 2nd ed., D, p. .40
*—— Pump fitter's guide, plates, D, paper .40
*—— The gas fitter's guide, D, paper, .40
‡ Engineers' Data book, illus., pap., S, 1.00
*Fahie, A. House Lighting by Electricity, O, .80
Fahie, J. J. History of electric telegraphy, D, 3.00
Fitzmaurice, M. Plate Girder Railway Bridges, O, 2.40
Fishbourne, G. Stability the Seamen's Safeguard, S, .40
Fleming, J. A. Short lectures to electrical artisans, D, 1.50
Fletcher, W. Steam Locomotion on common roads, O, 3.00
Foden, J. Boiler-maker's companion, D, 2.00
‡ Foster, J. Evaporation by the multiple system. illus. O. 7.50
French polisher's manual, Tt, paper. .20
Fullerton, W. Architectural examples, 200 plates O, 6.00

SCIENTIFIC BOOKS.

George, E. M. Pocketb'k of Calculations in Stresses, Tt, 1.50
Gillett, W. The phonograph and how to construct it, D, 2.00
Girder, W. J. Weight of iron, folding card. .40
*Gorham, J. Construction of crystal models, plates, D, 2.00
Graham, D. A. Commercial values of gas coals, O, 3.00
Graham, J. C. Steam and the use of the indicator, O, 3.50
—— J. Elementary text-book on the calculus, D, *In press.*
——, M. Construc. and Work. Regenerator Furnaces, S, 1.25
Grant, J. Strength of cement, O, 4.25
Greenwell, G. C. Mine engineering. 3rd. ed. 64 plates Q 6.00
Grimshaw, H. Kitchen boiler and water pipes, O, .40
Gripper, C. Tunnelling in heavy ground, O, 3.00
Grover, J. W. Estimates etc. for railway bridges, F, 12.50
—— Iron and timber railway superstructures, F, 17.00
Haldane, J. W. C. Civil and mech. engineering, O, 4.50
—— Steamships and their machinery, O, 6.00
Hallatt, G. W. T. Hints on arch. draughtsmanship, T, .60
Halliday, G. Mechanical drawing. 12 plates, F, 2 pts., each .75
*—— Notes on design of small dynamos, O, 1.00
—— Mechanical graphics, O, 2.00
—— Belt Driving, O, 1.50
‖Handy Sketching Book, 5in. x 8in., paper, .25
‖———————— Pad, 10in. x 8in., paper, .25
Hansen, E. C. A Manual on Fermentation Industries
    for Brewers, Distillers, Wine Manufacturers, O, 5.00
‡Hardaway, B. H. Tables and formulæ for R. R. eng'rs. 2.00
Heaford, A. S. Strains on braced iron arches, O, 2.40
Heath, A. H. Manual on lime and cement, D, 2.50
Hedges, K. Continental Electric Light Central Stations, Q, 6.00
—— American Electric Street Railways, Q, 5.00
Hennell, T. Hydraulic tables, D, 1.50
Henthorn, J. T. and Thurber, C. D. The Corliss engine, S, 1.00
Hering, C. Winding magnets for dynamos, D, 1.25
Hett, C. L. Turbine manual and millright hndbk, O, pap, .80
Hick, J. Leather collars in hydraulic presses, O, pap. .40
‡ Higgs, P. Algebra self-taught, fourth ed., D, .60
H. L. Screw cutting tables for engineers, O, bds. .40
Hodgetts, E. A. B. Liquid fuel, O, 2.50
Hodgson, F. T. The Hardwood finisher, D, 1.00
Holloway, Thos. Levelling, O, 2.00
Hood, Chas. Warming buildings by hot air, etc. O, 6.00
Hoskiær, V. Testing telegraph cables, D, 1.50
Hoskins, G. G. The clerk of works, 3rd ed., D, .60
Hospitalier, E. Domestic electricity for amateurs, O, 2.50
Hornby, J. Gas Engineers Laboratory Handbook, D, 2.50
Hovgaard, G. W, Submarine boats, plates, D, 2.00
Hughes, N. Magneto Hand Telephone, S, 1.00

Hurst, Tredgold's elementary principles of carpentry, O,	5.00
Hutchinson, E. Girder making, plates, O,	4.25
*Jeans, J. S. Waterways and water transport, O,	5.50
*Iron and Steel Institute. Journal of the. Half-yearly, O,	6.00
* —— Proceedings in America *(special vol.)*, O,	6.00
Jenkin, F. Report committee electrical standards, O.	3.75
Johnson, F. R. Girder and Roof Trusses, D,	2.50
Jordan, C. H Weights of iron and steel, 4th ed., Tt.	1.00
—— Particulars of Dry Docks on the Thames, S,	1.00
Keerayeff, P. Tables of speeds, tr. by S. Kern, T, paper	.20
Kempe, H. R. Handbook of electrical testing, O,	7.25
Kent, W. G. The water meter, illus., D,	1.50
Kennedy, A. and Hackwood, R. W. Railway curves, Tt,	1.00
‡Kinealy, J. H. Steam Engines and Boilers, O,	2 00
* King, W., and Pope, T. A. Gold, Copper and Lead, D,	4.00
*Kirkpatrick, T. S. G. Hydraulic Gold Miners' Manual,	*press*
*—— Simple Rules for the Determination of Gems, S,	.80
Knight, C. Construction and manipulation of tools, Q,	7.25
Kutter, W. R. Hydraulics, tr. by L. D'A. Jackson, O,	5.00
La Nicca, J. Turners' and Fitters' pocket-book, pap.	.20
Lathes and Turning, Examples of lathes. O.	1.00
Laxton's Bricklayer's tables, Q,	2.00
—— Excavators tables, Q,	2.00
Leaning, J. Quantity surveying, 2nd ed., plates, D,	3.50
*Leask, A R. Triple and quadruple engines, &c., D,	2.00
*—— Breakdowns at sea and how to repair them, D,	2.00
*—— Refrigerating Machinery, D,	2.00
Lee, D. Manual for gas engineering students, S,	.40
Lindsay, Lord. Screw cutting tables for engineers, O,	.80
Little, G. H. Marine transport of petroleum, D,	3.50
Livingstone, D. Setting out of railway curves, D,	4.25
Lock, C. G. W. Sugar growing and refining. 200 illus., O,	10.00
—— Coffee: culture and commerce, illus., D,	4.00
—— Tobacco growing and manufacturing, illus., D,	3.00
—— Practical gold mining, illus. O,	15.00
—— Ore dressing machinery, Q.	10.00
—— Miner's pocketbook, limp leather, D,	5.00
‡—— Economic Mining, O,	5.00
*Longridge, J. A. The construction of ordnance, O,	10.00
*—— Internal Ballistics. O,	7.20
*—— Smokeless powder, O,	1.20
*—— The artillery of the future, O,	2.00
*Love, H. D Hydraulics, O,	2.00
Lovibond, T. W. Brewing with raw grain, O,	2.00
Luard, C. E. Stone: how to get it and how to use it, O,	.80
*Lukin, J. Turning lathes, illus., D,	1.00
Macfarlane, J. W. Pipe founding, plates, O,	4.00
Mackesy, W. H. Table of barometrical heights, Tt,	1.25
*Mcdermot and Duffield Treatment of gold ores, O,	2 00

Manning, R. Sanitary works abroad, O, paper,	.80
Mansergh, J. Thirlmere water scheme, maps, O, paper	.60
*Marshall, L. C. Practical flax spinner, illus., O,	6.00
Martin, W. A. Screw Cutting Tables, O,	.40
Masters, H. An architects letter, O, paper,	.40
*Matheson, E. Engineering enterprise abroad. illus., O,	
—— Depreciation of factories, 2nd ed., O,	3.00
—— and Grant. Handbook for engineers, pap., Tt.,	.80
Maxwell and Tuke. Disposal of sewage, O, paper	.40
*Maycock, W. P. Electrical notes, Tt,	.75
Merrett, H. S. Surveying, 4th ed., rev. by G. W. Usill.	5.00
Middleton, R. E. Measurements at the Forth bridge, O	1.20
Millar, W. J. Principles of mechanics, D	.60
Millis, C. T. Metal plate work, illus., D,	3.50
Molesworth, G. L. Metrical tables, Tt,	.60
*—— Pocket-book for civil and mech. engineers, Tt,	2.00
—— and Hurst, J. T. Pocket-book of pocket-books, Tt,	5.00
Moritz, E. R., Morris, G. H Science of Brewing, plates, D,	7.50
Murgue, D. Centrifugal ventilating machines, tr. O,	2.00
Myers, W. B. The "Schwedler bridge," plates, O, pap.	1.00
‡Nares, Capt. Sir G. S. Seamanship, plates, O,	3.00
Nelthropp, H. L. Watchwork: past and present, D,	2.50
Newman, J. Notes on concrete, 2nd edition, D,	2.50
Newman, J. Earthslips and subsidences, D,	3.00
—— Notes on Cylinder Bridge-Piers O,	2.50
*—— Scamping Tricks. D,	1.00
Nissenson, G India rubber manufacture, D, paper,	.75
—— Treatise on injectors, D, paper,	.50
Nystrom, J. W. Steam Engineering, O,	1.50
—— Elements of Mechanics, O,	2.00
Olander, E. New method of graphic statics, F,	4.25
Ornamental Penman's P'kt-bk. of Alphabets, D, paper,	.20
‡Ott, Karl von. Graphic statics. Tr. by G. S. Clarke, D,	1.50
*Paterson, M. M. Testing pipes and pipe-joints, O, pap.,	.80
Penman, W Land surveying, O,	3.50
Phillips, J. Drainage of towns, O, paper	.60
Porter, C. T. Richards Steam engine indicator. O,	3.00
Practical Electrics. Illustrated, D.	.75
Pray, T. J. Twenty years with the indicator. O.	2.50
‡ —— Steam tables and engine constants, O,	2.00
Price, W. Turners' handbook on screw-cutting. S.	.40
*Proc. Munic. and Sanitary engineers and surveyors, edited by Thos. Cole. *Published annually.*	
Procter, H. R. Tanning and manufact. leather. *In press.*	
Rapier, R. C. Remun. railways for new countries, Q.	6.00
Redwood, I. I. Practical Ammonia Refrigeration, S,	1.00
*Reed's Engineers' h'nd'k, by W. H. Thorn. 13th ed., O,	4.50
Reeves, R. H. Bad drains and how to test them, D,	1.40
Reid, H. Manufacture of portland cement, plates, O,	
Reis, P. Inventor of the telephone, by S. P. Thompson.	3.00

Reynier, E. Voltaic accumulator, Tr, O,	3.00
Richards, W. Gas consumer's handy book, S, paper	.20
—— Manufacture of coal gas, plates, Q,	12.00
‡Richards, J. Operation of woodworking factories, D,	1.50
—— Workshop manipulation, D,	
‡ —— Centrifugal Pumps, O,	1.00
Richmond, G. Gas and oil engines, Tt,	1.00
Rigg, A. Treatise on the steam engine, plates, Q,	10.00
*Rigg, A. and Garvie, J. Modern guns and smokeless powder, O,	2.00
Ritter, Prof. Iron roof and bridge construction, tr. by H. R. Sankey, O,	6.00
*Roberts, Charles W. Practical Advice for Marine Engineers, S,	1.00
Robertson, F. Engineering notes, O,	5.00
Robinson, H. Sewage disposal, 2nd ed., D,	2.00
—— Gas and petroleum engines. *New edition in preparation*	
*Robinson, H. Systems of distributing electricity, O, pap,	.80
Routledge, T. Bamboo as a Paper-making material, pap.	.80
Rowan, T. Disease and putrescent air, O, paper	.80
—— Spontaneous combustion of coal, O,	2.00
Rowell, H. Manual of hard soldering, D,	.75
Salis, R. de Hydraulic tables for circular sewers, O, p.	.40
Salwey, E. R. Light railways. O,	2.00
Sang, E. Lessons in applied science. 3 pts, D, each	1.25
Scamell, G. Breweries and malting. Second edition, by F. Colyer, plates, O,	5.00
*Saunders, C. A. Handbook of Practical Mechanics, S,	1.00
*Screws and Screw Making. D,	1.25
Sexton, M. J. Pocket-book for boiler-makers, 2nd ed.	2.00
Shaw, E. M. Fires in theatres, new edition D.	1.25
Simmonds, P. L. Useful animals and their products, S,	.80
—— Animal food resources of different nations, D,	1.00
‡—— Hops: cultivation, commerce, and uses, D,	1.25
—— Tropical Agriculture, O,	8.00
Smeaton, J. Plumbing, drainage, hot water fitting, D,	3.00
*Society of Engineers transactions for 1890, O,	6.00
Spang, H. W Lightning protection, D,	.75
Spencer, A. Roll turning, 56 large folding plates, O,	8.00
—— Appendix to Roll turning, 22 folding plates, O,	4.25
—— Ditto, complete in 2 vols., 78 plates, O,	12.00
Spons' Dictionary of Engineering. In 8 divs. cloth $5.00 each; 3 vols. cloth $40.00; 3 vols. half morocco	50.00
—— ditto Supplement. In 3 divs. cloth $5.00 each; in 1 vol. cloth $15.00 ; in 1 vol. half morocco	20.00
Spon, E. Sinking and boring wells, 2nd ed., D,	3.50
Spons' Encyclopædia, 5 v. cloth $27, 2 v. half morocco	35.00
—— Mechanics' own book, illus., O,	2.50
—— Tables and memoranda for engineers, 8s, (*in case*)	.50
Spretson, R. E. Casting and founding, 5th ed., plates, O,	6.00

Sprague, J. T. Electricity, theory and practice, 3rd ed, D.	6.00
Standage, H. C. Polish and varnish maker, D,	2.50
*Stanley, W. F. Motions of fluids, illus, O,	6.00
*Stanley, W. F. Surveying and Levelling Instruments, D,	3.00
*—— Mathematical drawing instruments. illus., D,	2.00
Steel, J. Malting and brewing, plates, O, 2 vols,	7.50
Stephens, V. Wrinkles in electric lighting, D,	1.00
Stone, T. W. Simple hydraulic formulæ, D,	1.50
*—— Water supply in new countries, D,	2.00
Stoney, B. B. Strength and proportion of riveted joints, O,	2.00
Streatfeild, F. W. Organic chemistry. D,	1.25
‡ Stuart, D. M. D. Coal Dust an Explosive Agent, Q.	3.00
Symons, G. J. Lightning rod conference, O,	3.00
Terry, G. Pigments, paint and painting, illus., D.	3.00
* Thompson, S. P. Electricity in mining, paper, O,	.80
‖—— Dynamo Electric Machinery, 5th edition, O,	5.50
‖—— Polyphase electric currents, O,	3.50
—— Electrical tables and memoranda, 8s,	.50
*—— Murcurial air-pump, illus., O, paper,	.60
—— The electromagnet, O,	6.00
‖—— Polyphase electric currents, illus., O,	3.50
Turner and Brightmore. Waterworks Engineering, O,	10.00
Turning. Geometrical turning simplified, O,	1.25
—— Examples of lathes, apparatus and work, O,	1.00
Unwin, W. C. Short logarithmic tables, O,	1.40
*Uppenborn, F. History of the transformer. D, paper.	.75
*Useful Hints to sea-going engineers, D,	1.40
Vosmaer, A. Iron and steel, D,	2.50
*Walker, S. P. Electric lighting for marine engineers D,	
Walmisley, A. T. Iron roofs, 2nd ed, plates, Q, hf. mor.	21.00
Walsh, M. Brickmaking in Western India, O, paper,	.40
Watson, E. P. How to run engines and boilers, S.	1.00
‡ Weightman, W. H. Position Diagram of Cylinder, with adjustable valves (Meyer's) printed on card,	.25
Welch, E. J. C. Designing belt gearing, S, paper,	.20
‡—— Designing slide valve gearing, D,	1.50
Wheeler, W. H. Drainage of low lands, plates, O,	4.00
—— Canals. O, paper,	.40
Willcocks, W. Egyptian irrigation. Illus. O,	15.00
Wood W. H. Stairbuilding and Handrailing, Q.	4.25
Woodward, C. J. Five Figure Logarithms, S.	1.60
Workshop Receipts. Mechanical, chemical, electrical, and metallurgical, five volumes, each	2.00
Wurtele, A. S. C. Standard measures, O,	.50
*Wylie, C. Treatise on iron founding, illus., D,	2.00
Young, W. Municipal Buildings Glasgow. F.	4.25
—— Town and country mansions. plates, Q.	12.50
Zeuner, Dr. G. Valve-gear, tr. by Prof. J. F. Klein, O,	5.00

LIST OF PERIODICALS

English

*Electrician	$9.00
*Electrical Review	6.50
*Engineering, thick paper, $10.00; thin paper	9.00
†The Engineering Review	1.75
*The Engineer, thick paper, $11.00; thin paper	9.00
†The Marine Engineer	1.75
†The Photogram	1.10
†The Process Photogram	2.00

American

*Architecture and Building	6.00
*American Gas Light Journal	3.00
†American Brewers' Review	5.00
*Brewers' Journal	5.00
*Electrical Engineer	3.00
†Electric Power	2.00
*Electrical World	3.00
†Journal of Electricity	1.00
†Journal of the Franklin Institute	5.00
†Locomotive Engineering	2.00
†Mining	1.00
†Machinery	1.00
†Popular Science	1.00
†Power	1.00
†The Safety Valve	1.00
†The Stationary Engineer	1.00
‡The Engineer	2.00
†The Engineering Magazine	3.00
*The Engineering and Mining Journal	5.00
*The Engineering News	5.00
†The Painters' Magazine	1.50
*The Railway Age	4.00
*The Railroad Gazette	4.20
*The Scientific American	3.00
*The Scientific American Supplement	5.00
*The Western Electrician	3.00

☞ * Weekly.   † Monthly.   ‡ Fortnightly.

The Photogram

A monthly magazine devoted to Photography and all its appurtenances, supplies, new materials, etc., fully illustrated and containing much valuable information and practical hints for the amateur or the professional. Small folio size, well printed on nice paper.

Annual subscription $1.10, payable in advance.

The Process Photogram

A monthly magazine. This magazine is composed of THE PHOTOGRAM and 8 to 14 additional pages on Process work, new methods, new appliances, information on printing and color work, and all the most advanced practice obtainable. A magazine that every process-artist ought to have on his shelves for reference. This is without doubt the most complete and up-to-date magazine ever done on this subject. Fully illustrated.

Annual subscription $2.00, payable in advance.

We also receive subscriptions for American and Foreign Engineering, Technical and Scientific Journals and Magazines. All subscriptions payable in advance by P. O. O., Express Money Order, Registered Letter, or Cheque or Bank Draft on New York, and made payable to

SPON & CHAMBERLAIN

12 Cortlandt St., N. Y.

THE MOST COMPLETE WORK ON THE SUBJECT.

---o---

CORLISS ENGINES,

and Allied Steam-Motors, working with and without automatic variable expansion gear.

A practical treatise on the development, progress and constructive principles of these engines for engineers, machinists, steam users and engineering colleges.

A translation of W. H. Uhland's work, with additions,

By ANATOLE TOLHAUSEN, C.E.

In two volumes :—

Vol. 1.—Text, 288 pages, 386 illustrations, 4 large pages of tabulated matter and 33 plates, 4to., half-morocco.

Vol. 2.—66 double page plates, large folio, half-morocco.

Complete in 2 vols. $21.00.

---:o:---

Slide and Piston-valve Geared Steam Engines,

Forming a supplement to Corliss Engines and Allied Steam-Motors.

Translated from W. H. Uhland's work, with additions,

By ANATOLE TOLHAUSEN, C.E.

In two volumes :—

Vol. 1.—Text, 154 pages, fully illustrated, and 20 large plates, 4to., half-morocco.

Vol. 2.—27 double page plates, large folio, half-morocco.

Complete in 2 vols. $14.00

THE CORLISS ENGINE.

By John T. Henthorn.

— AND —

MANAGEMENT OF THE CORLISS ENGINE.

By Charles D. Thurber.

Uniform in One Volume, Cloth Cover; Price, $1.00.

Table of Contents.

Chapter I.—Introductory and Historical; Steam Jacketing. Chapter II.—Indicator Cards. Chapter III.—Indicator Cards continued; the Governor. Chapter IV.—Valve Gear and Eccentric; Valve Setting. Chapter V.—Valve Setting continued, with diagrams of same; Table for laps of Steam Valve. Chapter VI.—Valve Setting continued. Chapter VII.—Lubrication with diagrams for same. Chapter VIII.—Discussion of the Air Pump and its Management. Chapter IX.—Care of Main Driving Gears; best Lubricator for same. Chapter X.—Heating of Mills by Exhaust Steam. Chapter XI.—Engine Foundations; diagrams and templets for same. Chapter XII—Foundations continued; Materials for same, etc.

Third Edition, with an Appendix.

Mailed post paid to any address in the world on receipt of $1.00.

Practical Handbooks.

Cromwell, J. H.—A System of Easy Lettering   50cts.

Dahlstrom, K. P.—The Fireman's Guide. A handbook on the care of boilers. 28 pages, cloth,   50cts.

Fawkes, F. A.—Hot Water Heating on the Low Pressure System. 78 pages, illustrated, boards,   40cts.

Graham, Maurice.—Practical Hints on the Construction and Working of Regenerator Furnaces. 131 pages, 50 illustrations, 16mo., limp leather,   1.25

Henthorn-Thurber.—The Corliss Engine and Its Management. A practical book for the engineer and those in charge of Corliss engines. 96 pages, illustrated, 16mo., cloth,   1.00

Higgs, W. P.—Algebra Self Taught. A valuable aid for Engineers preparing for Examination,   .60

Hornby, J.—The Gas Engineer's Laboratory Handbook. 304 pages, 83 illustrations, 12mo, cloth,   2.50

Hughes, N.—The Magneto Hand Telephone: its construction, fitting up and use. 80 pages, 23 illustrations, 12mo, cloth,   1.00

Ornamental Penmans Pocket Book of Alphabets, for Engravers, Sign Writers and Stone Cutters   20cts.

Practical Electrics.—A universal handy book on everyday electrical matters. 135 pages, fully illustrated, 8vo, cloth,   75cts.

Rowell, H.—Manual of Instruction in Hard Soldering. 56 pages, illustrated, cloth,   75cts.

The Handy Sketching Book: for the use of engineers and draughtsmen. Sectional to EXACT eighths of an inch, blue lines, stiff boards, with useful tables, 25 cents each. Per dozen, $2.50.

Watson, E. P.—How to Run Engines and Boilers. Practical Instructions for Young Engineers and Steam Users. 125 pages, illustrated, 16mo., cloth,   1.00

Weightman, W. H.—Position Diagram of Cylinder, with (Meyer) Cut-off at one-eighth, one-quarter, three-eighths and one-half; with adjustable valves. Size 6in. x 8in., printed on card,   25cts,

Workshop Receipts. A valuable technical series in five

NEW BOOKS.

Dynamo Electric Machinery. A manual for students of electrotechnics. By Silvanus P. Thompson, D.Sc. **5th** edition, thoroughly revised and rewritten, new illustrations and folding plates. 832 pages, 540 illustrations, 8vo, cloth... $5.50

Polyphase Electric Currents, and alternate current motors. By S. P. Thompson, D.Sc. The best work on this subject. 250 pages, 171 illustrations, 8vo, cloth.................. 3.50

Appleby's Handbook. Machine and Hand Tools... 1.50

Economic Mining. A practical handbook, including latest practice. By C. G. Warnford Lock, M.I.M.E. 668 pages, 175 illustrations, 8vo, cloth............................. 5.00

The Manufacture of Soap and Candles, Lubricants, Glycerine. By W. Lant Carpenter. Second edition revised by Henry Leask. 446 pages, 104 illustrations, 12mo, cloth..... 4.00

An Elementary Text-book on Steam Engines and Boilers. By J. H. Kinealy. AMERICAN PRACTICE. 236 pages, 103 illustrations, 8vo, cloth... 2.00

THE BEST BOOK ON ICE MAKING.

Theoretical and Practical Ammonia Refrigeration. By I. I. Redwood, M.E. AMERICAN PRACTICE. A practical and thoroughly reliable handbook for the engineer and all those engaged in the running of ice and ammonia refrigerating machinery. 156 pages, 15 illustrations, 24 pages of tables, 12mo, cloth....................................... 1.00

The Management of Gas and Oil Engines, including a full description of the Hornsby-Akroyd oil engine. By G. Lieckfeld, C. E. A practical handbook for the instruction of those using gas engines and oil motors. Fully illustrated, 12mo, cloth............... 1.00

CLEANING AND SCOURING:

A MANUAL FOR DYERS AND LAUNDRESSES AND FOR DOMESTIC USE.

BY S. CHRISTOPHER.

CONTENTS.

Dresses.—Silk, Satin, Irish Poplin and Tabinet, Lama, Alpaca, Printed Muslin and Pique, Pique and Colored Muslin.

Shawls and Scarves.—China Crape, Brocaded or Printed Silk, and Woolen.

Silk Handkerchiefs, Ribbons, Mantles, Fancy Waistcoats, and Lace.

Gloves.—Kid, Washleather.

Feathers.—White, Colored, Grebe; to purify—for Beds, Pillows, &c.

Bonnets.—Chip, Straw, and Leghorn.

Ancient Tapestry.

Curtains, Bed Furniture, &c.—Chintz, Damask, Worsted-and-cotton Damask, French Damask—Silk-and-worsted Moreen, Tabaret or Tabbarea, Satin, Tammy Lining, Fringes—Bullion and worsted, Lace and Gimp—Bullion.

Table Covers.—Silk-and-worsted, Cotton-and-worsted, and Printed Cloth.

Carpets.—Dry Cleaning, thorough Cleaning.

Hearthrugs, Sheepskin Rugs and Mats.

To Remove Various Stains from Linen and Cotton.—Fruit Stains, Grease Spots, Ink Stains, Marking Ink, Mildew, Paint or Varnish, Wine Stains.

Recipes for general Domestic Use.—Oilcloth, Paint. Floors, Marble, Iron and Steel, Brass or Copper, Silver Plate, Furniture, Gilt Frames, Ivory Ornaments, Mirrors, Wall-paper, Stone Steps.

Definitions, &c.—Boards, &c., for French Cleaning, Camphine, Common Sour, Drying, Frame, French Board, Hot Stove, Irons, Parchment Size, Pegs, Puncher, Size, Soap,

The Engineer.

A FORTNIGHTLY PAPER FOR ENGINEERS, STEAM USERS, AND ALLIED TRADES.

THIRTEENTH YEAR.

Contains useful matter of the class indicated by this work, and treats all subjects in a plain matter of fact way, without pedantry and without pretense.

Two Dollars per Year Always in Advance.

EGBERT P. WATSON & SON,

150 Nassau Street,   New York.